Microwave Assisted Chemistry Experiments

(Organic Synthesis, Chemical Analysis and Extraction)

Microwave Assisted Chemistry Experiments

(Organic Synthesis, Chemical Analysis and Extraction)

B.R. Prashantha Kumar
M. Pharm, PhD
Assistant Professor, Dept. of Pharmaceutical Chemistry,
JSS College of Pharmacy, Mysuru, India

T. Durai Ananda Kumar
M. Pharm, PhD
Assistant Professor, Dept. of Pharmaceutical Chemistry,
JSS College of Pharmacy, Mysuru, India

N. Swathi
M. Pharm, PhD
Associate Professor, Dept. of Pharmaceutical Chemistry,
Gokaraju Rangaraju College of Pharmacy, Hyderabad, India

PharmaMed Press
An imprint of BSP Books Pvt. Ltd.
4-4-309/316, Giriraj Lane,
Sultan Bazar, Hyderabad - 500 095.

Microwave Assisted Chemistry Experiments
(Organic Synthesis, Chemical Analysis and Extraction)
***by* B.R. Prashantha Kumar, T. Durai Ananda Kumar and N. Swathi**

Published by

PharmaMed Press
An imprint of BSP Books Pvt. Ltd.
4-4-309/316, Giriraj Lane, Sultan Bazar, Hyderabad - 500 095.
Phone: 040-23445688; Fax: 91+40-23445611
e-mail: info@pharmamedpress.com
www.pharmamedpress.com/pharmamedpress.net

ISBN: 978-93-89974-95-9

FOREWORD

Microwave technology for chemical synthesis has come to an age after two decades of presence in the tool box of organic chemists. It definitely improves the energy efficiency and reduces the reaction time, hence follows the green chemistry principles. This emerging field of microwave chemistry is adventurous and rewarding with respect to time and efficiency.

A resource material for the basic and fundamental knowledge is essential. With the interest of fulfilling the above mentioned, the authors Dr. B. R. Prashantha Kumar, Dr. T. Durai Ananda Kumar and Dr. N. Swathi have attempted to construct this book titled "Microwave Assisted Chemistry Experiments (*Organic Synthesis, Chemical Analysis and Extraction*)" for the academic fraternity of chemistry especially organic chemistry, pharmaceutical chemistry and natural product chemistry.

This book features important aspects of microwave chemistry including the principle, mechanism and procedures for the chemical synthesis and their analysis. This book gives easy access to important reactions and practically feasible synthetic path for reactions which opens the scope for further investigation and innovation making this technique available for diverse applications. Inquisitively, authors have compared conventional and microwave methods for all the experiments.

Altogether, I strongly recommend this book cum practical manual to the students, research scholars and teaching fraternity of various chemistry disciplines.

Dr. Laxmi Adhikary
Chief Scientific Officer
Mabxience Insud Pharma
Madrid, Spain

PREFACE

We the authors submit our salutations to His Holiness Jagadguru Sri Shivarathri Deshikendra Mahaswamiji of Jagadguru Sri Veerasimhasan Math, Sutturu, Karnataka and present this book titled, "Microwave Assisted Chemistry Experiments (*Organic Synthesis, Chemical Analysis and Extraction*)" with his blessings.

Microwave chemistry is a unique and most adventurous technique, which offers efficient synthesis of active pharmaceutical ingredients (APIs) and drug intermediates with high reproducibility. Microwave technology converted several hazardous chemical transformations into non-toxic synthesis in an energy efficient way. In order to provide the information about the Microwave chemistry, we spent more than a decade to construct this book. All these experiments are optimized in the Pharmaceutical Chemistry laboratories of JSS College of Pharmacy, Ootacamund, Tamilnadu, JSS College of Pharmacy, Mysuru, Karnataka and Gokaraju Rangaraju College of Pharmacy, Hyderabad, Telangana. These reactions are verified several times by conducting experiments for the students. In recognition, to the knowledge and expertise in the field we published 12 research articles in peer reviewed journals.

This book reports microwave assisted chemistry experiments in a comprehensive and very simple way covering the principles and applications of microwaves in the field of organic synthesis, organic analysis and natural product extraction. The book imparts equal importance to the principle, reaction mechanism and experimental procedures. The reaction mechanisms component of each experiment adds real interest to revise key aspects of organic chemistry related to microwaves. The nomenclature used conforms as far as possible with Chemical Nomenclature.

The book comprises of five components or parts. The part 1, contains the basic introduction along with the important applications of microwaves. The part 2 is built with all the important organic synthesis (prescribed by several academia or universities), theory, principles and procedures. Qualitative and quantitative analysis of organic compounds is given in the part 3 and part 4, respectively. The part 5 includes chemical experiments related to extraction and isolation. We authors strongly believe that this book serve as a resource for undergraduate and post graduate students of chemistry (B. Sc and M. Sc) and also for pharmaceutical chemistry graduates (B. Pharm and M. Pharm). This book also serves as a reference guide for the research scholars in academia.

The heart of microwave ovens and reactors happens to be the Magnetron, which generates microwaves. In the market, variety of microwave ovens and reactors are available with different Magnetron capacities. In this backdrop, we authors, suggest the readers to set the power/intensity and time accordingly to suit your equipment and requirement. The basic understanding remains same and optimizing reactions for the specific microwave instrument that you have provides better results.

The author's thanks are due to Dr. Shrisaillappa Badami, M. Pharm, PhD, Former Director, Radiant Research, Bengaluru, for kindling interest to write this book way back in the year 2007.

Dr. B. R. Prashantha Kumar would like to thank Dr. Laxmi Adhikary and Dr. B. R. Srinivasa, Former Scientists at Astra Zeneca Research Foundation, Malleshwaram, Bengaluru, for their guidance and support to extend the facilities for microwave based chemistry experiments.

Our sincere gratitude to the Principals of JSS College of Pharmacy, Ootacamund, Tamilnadu, JSS College of Pharmacy, Mysuru, Karnataka and Gokaraju Rangaraju College of Pharmacy, Hyderabad, Telangana. Our heartfelt thanks to Mr. G. S. Mahendra and Dr. Gyanendra Kumar Sharma for their help to cross check the experiments for their reproducibility in different laboratories.

We also extend our special thanks to Mr. Anil Shah, Director, PharmaMed Press, Hyderabad., and Mr. Naresh Daveragave and Miss. Rajitha for their support in this project.

The love and understanding of our family members enabled us to complete this book successfully, and we acknowledge them.

Adequate measures are taken to print error free material. However, if anything is found, kindly communicate your suggestions and critical points.

- Authors

CONTENTS

PART - 1

INTRODUCTION

Microwave-Assisted Organic Reactions

Green chemistry can be defined as the design, manufacture and use of efficient, effective, safe and environmental friendly chemical processes and products. Microwave-assisted organic reactions cover all the principles of green chemistry. Microwave assisted organic reactions are useful in the synthesis of active pharmaceutical ingredients (APIs), drug intermediate and other compounds with chemical and medicinal importance (analytical, diagnostic, research). This technology improves the chemical process and reduces the pollution (solvent free methods). Microwave-assisted reactions maximises the efficient use of safer raw materials and reduces the waste (toxic material) generation.

1) ***Speed***: Microwave reactions can be completed in minutes. Some chemical reactions complete in seconds. In many cases, it reduces the reaction time from hours to minutes to seconds.
2) ***Economy***: Microwave reactions utilize no or low volume of solvents.
3) ***Cost effective***: Microwave reactions reduce the cost per microwave reactions mainly through increasing the reaction rate there by yields.
4) ***Simplicity***: The products of microwave reactions can be isolated very easily and requires no purification (recrystallization) in most cases.
5) ***Consistency***: Microwave reactions are reproducible.
6) ***Rapid optimization***: Microwave reactions complete very fast. Hence, the organic reaction optimization can be achieved faster than the conventional synthesis.
7) ***Energy efficient reaction***: Microwave reactions offer enhanced reaction conditions.

8) ***Higher yield***: The rapid-efficient reaction inhibits the byproducts formation and hence offers higher yields of the products.

9) ***High purity:*** The rapid-efficient reaction inhibits the byproducts formation and hence offers highly pure compounds.

10) ***Superheating:*** It takes the reaction environment to very high temperature (super heating). It is very essential for the several reactions such as substitution and coupling reactions.

11) ***Versatility:*** The microwave heating can be utilized for all kinds of organic reactions. It includes substitution, coupling, rearrangement, oxidation and reduction, etc.

Microwave Heating

Electromagnetic waves frequency ranges between 300 MHz and 300 GHz are named as microwaves. Most of the microwave ovens and microwave processors operate at 2.45 GHz (~12.2 cm λ). These microwaves penetrate into fogs and clouds and travel in straight lines. Microwave radiations were utilized in the development of Radio Detection and Ranging (RADAR).

In 1946, the American electrical engineer Dr Percy Spenser, noticed the melting of candy bar placed in his pocket under the exposure to microwave radiation once the magnetron was switched on. He was engaged in the experiments to utilize the magnetron in RADAR. This observation stimulated him to develop microwave oven. Based on this, he applied the magnetron heat for cooking popcorn and found working. This is stimuli for the development of the most popular and useful microwave oven in 1970. Initially microwave heating was utilized for heating water, moisture analysis and wet ashing procedures in chemical and biological laboratories. The computerized microwave ovens were used for the acid digestion of ores and minerals. Gedey *et al* and Giguere *et al* (1986) demonstrated the use of microwave ovens in organic reactions for the first time.

Theory of microwave heating: The rotational states of the molecules undergo excitation with electromagnetic radiation. The microwave irradiation, when absorbed by organic molecules induces the rotational changes. The frequency of molecular rotation is similar to the frequency of microwave radiation. The molecule continually attempts to realign itself with the applied electric field and absorbs the energy. This effect is utilized in microwave ovens to heat food materials. Chemists also utilize the microwave irradiation as an energy source for chemical reactions.

- Microwave oven contains microwave generator called as magnetron (inside the string metal box). It receives electricity and converts them into high-energy radio waves.
- Microwave guide (channel) introduces microwave heat energy (radiation) into the heat compartment.
- The microwaves bounce back and forth off the reflective metal walls of the heat compartment.
- The microwaves penetrate the material to be heated (reaction vessel) and vibrate them to cause molecular friction. The rate of vibration decides the heating and initiates the reaction.

Principles

Microwave ovens more efficiently channel heat energy into the molecules. In the microwave heating process energy transfer occurs by three mechanisms namely dipole rotation, ionic conduction and interfacial polarization. Microwave ovens inject the energy directly into the molecules, rather than warming the outside walls of a reaction vessel to spread heat by convection and conduction. High frequency electromagnetic radiations (electric fields) exert a force on charged particles of molecules and that causes molecular friction to generate super heat.

- ***Ionic conduction***: Ionic conduction is the electrophoretic migration of ions, when an electromagnetic field is applied. The oscillating electromagnetic field generates an oscillation of electrons in a conduction and results electric current.

The conduction mechanism generates heat through resistance friction to the electric current.

- ***Dipole rotation***: It means rearrangement of dipoles with the applied field. Polar molecules are the ideal material for dipolar polarization. Dipole polarization depends on the dipole moment of a molecule. The difference in the electro negativity of the atoms and molecular symmetry is responsible for this effect. The alignment of polar molecules with an oscillating electromagnetic field results random motion of particles. This random motion effect generates heat. The dielectric polarization provides the energy to the molecules to rotate into alignment. The polarizations (Maxwell-Wagner effect) contribute heating effect.
- ***Interfacial polarization***: A combination of the conduction and dipole polarization mechanism.

How Microwave Radiations are Irradiated? Microwaves are heterogeneously distributed within the cavity and produces defined regions of high and low energy intensity. The energy variation can be minimized by smoothing mechanism, which disperses the incoming energy through a wave stirrer (mode stirrer). It is a reflective, fan-shaped paddle attached to the opening of wave-guide feed. The turn table (rotating platform) present in the microwave oven ensures that an average energy field experienced by the sample is approximately the same in all directions.

Superheating: Superheating of liquids is common under microwave irradiation because of molecular friction. In super heating, a liquid attains a temperature much above its conventional reflux boiling point. This superheating, which is not commonly seen in conventional heating may help in increasing the rate of reaction. Microwave irradiation provides superheating. Superheating in closed vessels and under pressure facilitates the organic reactions. Superheating offers highly accelerated reaction rate and enables chemical synthesis in much lesser time with good yields. Superheating of liquids or solutions under microwave irradiation raises the temperature above the conventional boiling point. It reduces several hour conventional reactions into fewer minutes microwave reactions. Water, for example reaches 105 °C (5 °C above actual boiling point) and acetonitrile reaches 120 °C (38 °C higher than normal boiling point).

Wall-heat transfer: The wall heat-transfer that occurs with heat resources such as water bath, oil bath and steam bath leads to incomplete reaction. Microwave irradiation produces efficient internal heat transfer (*in situ* heating), and overcomes the wall-heat transfer mechanism. The microwave heating reduces the tendency for seed formation (initiation of boiling).

Microwave Instrument Components

1) ***Magnetron***: Microwave oven magnetron converts the shortest microwaves (12 cm; 4.7 inches), which carry higher energy into electromagnetic radiations. A magnetron is a microwave source (thermionic diode) consist an anode and a cathode. Cathode releases electrons upon direct heating. The anode consists even numbered small cavities (tuned circuit). The gap across the end of each cavity behaves as a capacitance. The electrons released from cathode are attracted towards the anode. It causes bending of the path of electrons, when they travel from cathode to anode. These deflected electrons pass through the cavity gaps and induces a small charge in the tuned circuit. This is responsible for oscillation of the cavity and microwave generation.
2) ***Wave-guide feed***: A wave-guide feed is a rectangular channel made from a metal sheet. The reflective walls of wave-guide feed allows the transmission of microwaves from the magnetron to the oven cavity.
3) ***Oven cavity***: It refers to the place in an oven for placing the material to be heated. It is usually made of glass or fiber material of metal with reflective surfaces. Reflective surfaces increase the oven efficiency and to prevent the hazardous leakage. A wire mesh door of the cavities also prevents the microwave leakage. The ovens are equipped with fans to remove hot air and vapors and prevents oven from getting heated upto higher temperatures.

Microwave Solvents

Solvents serves a energy transfer media and help in coupling the thermal energy with the kinetic energy of the reactants. Solvents are of major concern as environmental pollutants (carcinogenic, mutagenic and allergens). Eco-friendly microwave chemistry requires no solvents or very lesser quantity of solvents as energy transfer medium. Rapid microwave synthesis leads to lesser evaporation of solvents and prevents or reduces environmental pollution. Polar solvents are best for dipolar polarization and microwave heating.

The solvents used in microwave reactions should possess dielectric heating property (Table 1). Dielectric heating ensures the conversion of electromagnetic energy into efficient heating. The ability of the solvent dielectric property is indicated by tan δ. The solvents with high tan δ provide rapid heating. However the solvent with low tan δ also can be used, but provide slow heating.

Table 1 Microwave solvents along with their dielectric constant values

High tan δ (>0.5)		Medium tan δ (0.1- 0.5)		Low tan δ (<0.1)	
Ethylene glycol	1.35	2-Butanol	0.45	Chloroform	0.09
Ethanol	0.94	Dichlorobenzene	0.28	Acetonitrile	0.06
DMSO	0.83	N-Methyl-2-pyrollidine	0.28	Ethylacetate	0.06
2-Propanol	0.80	Acetic acid	0.17	Acetone	0.05
Formic acid	0.72	DMF	0.16	THF	0.05
Methanol	0.70	Dichloroethane	0.13	DCM	0.04
Nitrobenzene	0.59	Water	0.12	Toluene	0.04
1-Butanol	0.57	Chlorobenzene	0.10	Hexane	0.02

Microwave Reaction Vessels

A safe reaction vessel for microwave heating for solvent mediated reflux reactions is essential. Microwave transparent materials such as Teflon, polystyrene, pyrex or borosilicate glass are useful in the vessel fabrication. These materials absorb the radiation poorly and with stand at higher temperatures.

High pressure increases the risk of explosion.

- ***Teflon***: Polyterafluroethylene (Teflon) offers resistant to strong bases and hydrogen fluoride. Longer, microwave exposure (more than 15 minutes) to Teflon materials softens the material, and may lead to loss of reaction content. Hence, the microwave reaction should be conducted in several pulses. Teflon is widely used material for preparation of sealed containers, which are commonly referred to as Teflon bomb.
- ***Nalgene***: It is an autoclavable and thermostable polypropylene material.
- ***Corian***: A durable and heat resistant polymer preferred for the organic reactions. This material permits the temperature rise to above 200 °C. It is desirable for prolonged reactions with microwave irradiation.
- ***Vermiculite***: It consists of hydrous silicates of ferrous, magnesium and aluminium. It is placed in either a Corian box, Nalgene dessicators or a container made of a special polymer.
- ***Glass wool***: It can be used as an alternative to vermiculite.

  A sealed reaction vessel (Teflon or Pyrex glass) covered with vermiculite absorb the reaction content in the event of explosion.

Open vessel reactions: Borosil beaker, conical flask and Erlenmeyer flask are useful. Glass wares covered with funnel and a watch glass avoids excessive solvent evaporation (incase of domestic microwave ovens). A flask fitted with condenser is available in microwave synthesisers.

Microwave-Assisted Chemical Reactions

1) **Dry media synthesis:** It is a most common microwave method. High pressure and associated danger of explosion can be avoided by dry media synthesis. It includes neat reaction and solid-support reactions.
    - ***Neat reaction***: It refers to reaction carried out without using solvent. A mixture of reactants without the use of solvent helps to avoid the risk of developing high pressure.
    - ***Solid-support reactions***: A reaction can be carried out by adsorbing the reactants on an inorganic solid supports under microwave irradiation. Inorganic solids namely in clay, silica, alumina and Zeolite are commonly useful solid supports (catalysts). The reactants adsorbed on the surface of inorganic solids absorb the microwave radiation. These solids prevent development of high pressure in sealed containers. In this kind of experiments, the reactant will be dissolved in organic solvent. Further, the reactant should be mixed with the inorganic solid support and followed by solvent evaporation. This facilitates the adsorption of reactants on the solid surface. Alternatively reactants can be triturated. After the microwave irradiation, the product can be extracted with suitable solvent.
2) **Solvent mediated synthesis:** High boiling polar solvents such as N,N-dimethyl formamide (DMF), *o*-dichlorobenzene, 1,2 dichloroethane (DCE) useful in the microwave reactions. Polar solvents with a high dielectric constant absorb microwave energy better than non-polar solvents due to dipole rotation. These solvents offers higher energy transfer rates. Water is an ideal solvent since it fulfils many criteria; non-toxic, non-inflammable and abundantly available and inexpensive. It possesses high polar character, novel-reactivites and selectivities. At higher temperature it behaves as a pseudo-organic solvent. The higher temperature decreases the dielectric constant and increases the solvating power comparable to ethanol and acetone.

    DMF and DCE are heated much faster than hexane or carbon tetrachloride in a microwave oven.

Ionic liquids: Salts (molten salts) in a liquid form at or below 100 °C. Ionic liquids absorb microwave irradiation extremely well and transfer energy quickly by ionic conduction. The major advantage of using ionic liquids in microwave-assisted organic reactions are:

1) Extremely low vapour pressure.
2) Thermal stability.
3) Electrochemical stability.
4) Non-flammability.
5) Catalytic property.

The instantaneous superheating of the ionic substance due to the ionic motion generated by the electric field. When the temperature increase, the transfer energy becomes more efficient.

Microwave-Assisted Extraction (MAE)

Extraction is one of the most important technique for the isolation, recovery and separation of phytoconstituents. The conventional extraction technique requires longer extraction time, large solvent volume and cause degradation to thermolabile substances. In contrast, the microwave-assisted extraction (MAE) requires less solvent volume, reduces the extraction time and prevents the oxidative degradation of phytoconstituents. Microwave extraction utilizes maceration and percolation priciples. The rate of extraction (breaking up of plants cells and tissues) in much higher in microwave extraction. In conventional extraction, the heat transfer occurs from heating medium to the sample. MAE offers heat directly to the medium (dissipation). The penetration of microwave energy depends on the dielectric property on the plant cell structure.

Safety Precautions

Microwave ovens are designed to ensure no leakage of radiation. The microwave ovens with high power capacity are useful. Even if there is any leaking, it is very much small in intensity when compared with cellular phone signals. This radiation leaking occurs at a distance of 5 cm (2 inch) and is about 5 MW per square cm. Microwave energy penetrates the skin into subcutaneous tissue and therefore also raises the temperature level of tissue and blood. Hence, the protection is required.

- ***United States-Food and Drug Administration* (US-FDA):** The United States-Food and Drug Administration (US-FDA) limits the leakage of microwave energy and ensures the safety. Shielding or increasing the distance from the oven ensures safety. In title 21 US-FDA states that the power density limit from an operating microwaves oven 'shall not exceed 1 mill watt per square centimetre at any point 5 centimetres or more from the external surface of the oven, measured prior to acquisition by a purchaser and thereafter, 5 milliwatts per square centimetre at any such point.
- ***World Health Organization* (WHO):** WHO warns that the thermal damage would occur from long exposures to very high power levels. In other words, the radiation is very low to cause damage to the tissue.

Advantages

1. Microwave irradiation penetrates the walls of the vessel and heats only the reactant and solvent, not the reaction vessel.
2. The energy transfer occurs through dielectric ions. In conventional methods, it occurs through conduction and or convection.
3. The commercial microwave systems (specific for synthesis) offer improved reproducibility, rapid synthesis and rapid optimization.
4. Microwave energy offers much higher heating rate (2 to 4 °C/s). In conventional heating it is very difficult to achieve this heating rate.
5. Microwave heating facilitates the solvent-free and solid-supported reactions (dry reactions).
6. The solvents heated under microwave energy, boil at elevated temperature. It increases the reaction temperature by upto 100 °C, above the boiling point. For example, ethanol (b.p 79 °C) provides temperature upto 164 °C (85 °C more).

7. In solid samples, the rate of energy transport is less and develops hotspots. These hotspots enhance the reaction rate as well as shift the equilibrium constant.
8. Microwave reactions offers higher efficiency for the multi-component reactions.
9. Microwave reactions promote the efficiency of combinatorial chemistry reactions.

Applications

1) Microwave irradiation is very much useful in the following chemical reactions.
   - Protection and deprotection reactions: Functional group protection and deprotection strategies are important in carbohydrate chemistry and peptide synthesis. Microwave chemistry is more useful in these reactions.
   - Named organic reactions: Gabriel synthesis, Suzuki reaction, Williamson-ether synthesis and Pinacol-Pinacolone rearrangement.
   - Oxidation, esterification, O-alkylation and aromatic electrophilic substitution reactions.
   - Preparation of medicinal compounds such as sildenafil, phenytoin, benzocaine are attempted successfully.
   - Drug intermediates, namely, 1,4-dihydropyridines, chalcones, carvones. Thioflavanoids and γ-carbolines also are synthesised.
   - Novel cepahalosporins are synthesised using microwaves.
   - Asymmetric reactions were also successfully attempted.
2) In analytical chemistry
   - Ashing: The ash content determination such as loss on ignition (LOI) and residue on ignition (ROI) is an important quality control procedure. Microwave ashing provides reduced time, cost and reduced exposure to fumes.

- Digestion: Microwave digestion assists in dissolving the metals in minutes during elemental analysis. Microwave digestion can oxidize compounds more effectively than conventional methods.
- Moisture analysis: The microwave-assisted moisture analysis overcomes the limitations of Karl-Fischer (destructive). Microwave analyzers works on higher dielectric constant (attentuate energy transfer) and offers non-destructive moisture measurement.

3) In natural product extraction and isolation: Extraction and isolation of active principles from plant materials is very critical. The conventional methods requires longer extraction procedures and higher solvent volume. Longer extraction and high volume of solvents pose the thermal degradation issues. Microwave-assisted extraction (MAE) is useful in the extraction of plant tissues using relatively less volume of solvent with higher extraction power. Thus it helps in overcoming the issues of conventional methods.
4) In food industry: Microwave heating is successfully applied in the food processing such as pasteurization and preservation. Microwave pasteurization is an efficient technique for milk and fruit juice. Microwave blanching is an another important application for the food industry. for freezing, canning and drying processes.

Important Microwave Reactions

Fewer important microwave reactions

1) **Microwave-assisted Wolf-Kishner reduction**

Wolf-Kishner reduction converts a carbonyl compounds (aldehyde or ketone) into an alkane using hydrazine hydrate and potassium hydroxide. The conventional reaction requires elevated temperatures and prolonged heating. Parquet and Lin (1997) reported the microwave assisted conversion of isatin into oxindole. Isatin was mixed with hydrazine and ethylene glycol and irradiated (microwave heating) for 30 seconds. This produced hydrazide product. The mixture of isatin hydrazide and potassium hydroxide in ethylene glycol was further irradiated for 10 seconds to obtain oxindole.

Isatin + NH_2NH_2 (Hydrazine hydrate) —MWI, 30 s→ Isatinhydrazide —KOH, Ethylene glycol, MWI, 10 s→ Oxindole

Ref: *J Chem Edu, 1997, 74(10), 1225*

2) **Microwave-assisted Aza-Prins cyclization**

Aza-Prins cyclization involves the condensation of homoallylic amine with an aldehyde in the presence of an acid to generate piperidines. Parchinsky et al (2011) reported the microwave assisted generation of 1-azaadmantanes (piperidines) from (4-methylcyclohex-3-ne-1-yl)-methyl amine. The microwave condensation of 1-(4-methylcyclohex-3-ne-1-yl)-methanamine with benzaldehyde (2 equivalent) in the presence of boron trifluoride diethyl etherate at 180 °C for 2 hours generated 4-methylidene-2,9-diphenyl-1-azaadamantane (83%).

1-(4-Methylcyclohex-3-en-1-yl)methanamine + 2 Benzaldehyde → ($BF_3.OEt_2$, MWI 180 °C, 2 h) → 4-Methylidene-2,9-diphenyl-1-azaadamantane

Ref: *Tetrahedron Lett, 2011, 52, 7161-7163.*

3) Microwave-assisted intermolecular Aldol condensation

Aldol condensation involves the condensation of enolate ion with a carbonyl compound in the presence of catalyst (acid/base) to generate β–hydroxy aldehyde (or β-hydroxy ketone). The dehydration of the β-hydroxy product forms conjugated enone. Molecules with dicarbonyl functionalities (e.g. 2-carboxybenzaldehyde) undergo special kind of aldol condensation called as intermolecular aldol condensation. This reaction constructs 5 or six membered rings. Koca et al (2018) reported the synthesis of 3-acetyl isocoumarin through intermolecular aldol condensation. The condensation of 2-carboxybenzaldehyde with mono-chloroacetone using triethyl amine as a base catalyst under microwave heating (300 W, 120 min) generated 3-acetylsisocoumarin.

2-Carboxy benzaldehyde + Chloroacetone → (MW 300 W, 120 min) → 3-Acetyl isocoumarin

Ref: *Arabian J Chem, 2018, 11,538-545*

4) Microwave-assisted Claisen condensation

Claisen condensation involves the condensation of two esters in the presence of strong base to generate β–ketoester. The condensation between one ester and another carbonyl compound also considered as claisen condensation (claisen reaction). Horta (2011) reported the synthesis diethyl-2, 5-dioxocyclohexane-1,4-dicarboxylate through intramolecular claisen condensation. The self condensation of diethyl succinate using *t*-butoxide as a base catalyst under microwave heating (150 °C, 20 min) generated the diethyl-2, 5-dioxocyclohexane-1,4-dicarboxylate (80%).

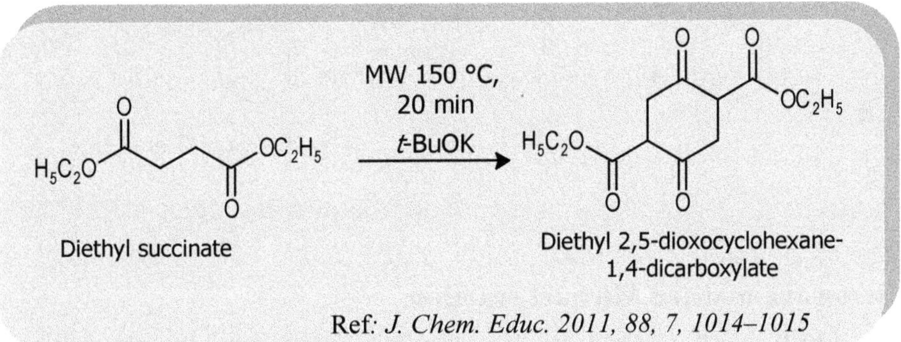

Ref: *J. Chem. Educ. 2011, 88, 7, 1014–1015*

5) Microwave-assisted Dieckmann condensation

Dieckmann condensation involves the intramolecular condensation of diesters containing carbon length of 6 and 7 in the presence of strong base to generate cyclic β–keto ester. Horta (2011) reported the synthesis ethyl 2-cyclopentane-carboxylate through Dieckmann condensation. The self condensation of diethyl adipate using *t*-butoxide as a base catalyst under microwave heating (150 C, 20 min) generated the 2-cyclopentanecarboxylate (~90%).

MW 150 °C, 20 min, *t*-BuOK

Diethyl succinate → Ethyl 2-oxocyclopentane-1-carboxylate

Ref: *J. Chem. Educ. 2011, 88, 7, 1014–1015*

6) Microwave-assisted Claisen-Schmidt reaction

The condensation of aromatic carbonyl compound (lacking α-hydrogen) with carbonyl compound (aldehyde or ketone) having α-hydrogen in the presence of base is known as Claisen-Schmidt reaction. This reaction generates α, β-unsaturated aldehyde or α, β-unsaturated ketone. Rayar et al (2015) reported the microwave condensation of benzaldehyde and acetone using sodium hydroxide to generate 4-phenylbut-3-en-2-one. The microwave heating of 5 W for 15 min offered the product.

CHO + H_3C–CO–CH_3 (Acetone) → MW 5 W, 15 min → 4-Phenylbut-3-en-2-one

Ref: *SpringerPlus, 2015, 4, 221-225*

7) Microwave-assisted Michael reaction

The addition of carbanion (or enolate, anamine; nucleophile) to the β-unsaturated systems (enone, enal) generates carbo-carbon bond at the β-carbon. This reaction is known a Micheal addition reaction. Zhao et al (2014) reported the microwave assisted Michael reaction for the synthesis of 1,3-diphenyl-3-[(pyridine-2-yl)-amino]prop-2-ene-1-one. The mixture of 2-amino pyridine and 1,3-diphenyl prop-2-en-1-one (chalcone) in ethanol upon microwave mediated heating (300 W, 30 min) produced 1,3-diphenyl-3-[(pyridine-2-yl)-amino]prop-2-ene-1-one (90%).

1,3-Diphenylprop-2-en-1-one + 2-Aminopyidine → MW 300 W, 30 min → 1,3-Diphenyl-3-[(pyridin-2-yl)amino] prop-2-en-1-one

Ref: *Res Chem Intermed, 2015, 41, 5809-5819*

8) Microwave-assisted Mannich reaction

The reaction between amine (1°, 2°, ammonia) and enolizable carbonyl compound (formaldehyde, methanal) produces β-amino carbonyl compound. This reaction is known as Mannich reaction. Aljohani et al (2019) reported the microwave assisted Mannich reaction for the synthesis of 1-{3-[(diethylamino)-methyl]-4-hydroxyphenyl}-ethan-1-one. The mixture of 4-hydroxybenzal-dehyde, formaldehyde and diethylamine upon microwave heating (300 W, 15 min) produced 1-{3-[(diethylamino)-methyl]-4-hydroxyphenyl}-ethan-1-one (77%)

4-Hydroxyacetophenone + Formaldehyde + $(C_2H_5)_2NH$ Dimethyl amine → (MW 300 W, 15 min) 1-{3-[(Diethylamino)methyl]-4-hydroxyphenyl}ethan-1-one

Ref: *Molecules, 2019, 24, 590-603*

9) Microwave-assisted Heck reaction (Mizoroki-Heck reaction)

The palladium catalysed cross-coupling reaction between alkenes and aryl (or vinyl) halides (or triflates) in presence of base to generate substituted alkenes is known as Heck reaction (one step synthesis of olifines). Ichikawa et al (2018) reported the Heck reaction for the synthesis of several alkenes. The cross-coupling reaction between the 3-iodoacetophenone and *n*-butylacrylate using tributylamine as a base catalyst and *N, N*-dimethylacetamide as a solvent under microwave heating (10 W) in flow chemistry principle generated butyl-3(3-acetyphenyl)acrylate (93%).

MW 10 W

3-Iodoacetophenone + *n*-Butylacrylate → Butyl 3-(3-acetylphenyl)acrylate

Ref: *Tetrahedron 2018, 74, 1810-1816*

10) Microwave-assisted Suzuki reaction (Suzuki-Miyaura reaction)

The palladium catalysed cross-coupling reaction between aryl (or vinyl) boronic acids with aryl (or vinyl) halide (or triflates) to generate alkenes, styrenes, biphenyls is known as Suzuki reaction. Dhara et al (2019) reported the Suzuki reaction for the synthesis of biphenyl compounds. The cross-coupling reaction between the iodobenzene and 4-methoxyboronic acid using palladium nanoparticles and under microwave heating (150 °C, 5 min) generated 4-methoxy-1,1'-biphenyl compound (95%).

MW 150 °C, 5 min

3-Iodobenzene + 4-Methoxyboronic acid → 4-Methoxy-1,1'-biphenyl

Ref: *Synth Commun, 2019, 49(6), 859-868.*

11) Microwave-assisted Niementowski cyclocondensation

The cyclocondensation reaction between anthranilic acid and amides generate quinazolin-4-ones. This reaction is known a Niementowski cyclocondensation. Alexandre et al (2002) reported the condensation of anthranilic acid and formamide under microwave irradiation (60 W, 20 min) to generate quinazolin-4-one (90%).

COOH, NH$_2$ + HCONH$_2$ —MW 60 W, 20 min→ Quinazoline-4-one

Anthranilic acid Formamide Quinazoline-4-one

Ref: *Tetrahedron Lett. 2002. 43. 3911-3923*

12) Microwave-assisted Biginelli reaction

A one pot condensation of aryl aldehyde, β-ketoester and urea (or thiourea) in acid catalysed reaction produces 3,4-dihydropyrimidine-1-ones (or 1-thiones). This reaction is known as Biginelli reaction. Jetti et al (2014) reported the synthesis of 3,4-dihydropyrimidine-1-ones (or 1-thiones) derivatives using microwave heating. The condensation of benzaldehyde, ethyl acetoacetate and urea under silica sulphuric acid catalysed microwave irradition (900 W, 7 min) produced 5-(ethoxycarbonyl)-6-methyl-4-phenyl-3,4-dihydropyrimidin-2(1H)-one (95%).

Benzaldehyde + Ethyl acetoacetate + Urea —MW 900 W, 7 min→ 5-Ethoxycarbonyl-(6-methyl-4-phenyl-3,4-dihydropyrimidin-2-(1*H*)-one

Ref: *Medicinal Chemistry Research, 2014, 23(10), 4356-4366*

13) Microwave-assisted Pinacol-Pinacolone rearrangement

The chemical conversion of 1,2-diols into carbonyl compounds through 1,2-migration reactions are known as Pinacol-Pinacolone rearrangement. Gutierrez et al (1989) reported the chemical conversion 2,3-dimethyl-butane-2, 3-diol (gem-diols, pinacol) into 3,3-dimethyl butan-2-one (pinacolone) using Al (III)-Monmorillonite K10 clay under microwave irradiation (450 W, 15 min). The reaction produces excellent yield (99%).

MW 450 W, 15 min

2,3-Dimethylbutane -2,3-diol → 3,3-Dimethylbutan-2-one

Ref: *Chem Soc Rev, 2012, 41, 1559-1584*

14) Microwave-assisted Beckmann Oxime rearrangement

The acid catalysed chemical conversion of oximes into amide (e.g. acetanilide) through selective migration of one group (rearrangement) is known as Beckmann oxime rearrangement reaction. Sugamoto et al (2011) reported the microwave assisted (350 W, 10 s) chemical conversion of acetophenone oxime into acetanilide (amide; 89%). This reaction is catalysed by the use of indium triflate and [Bdimim][PF6].

MW 350 W, 10 s

Acetophenone oxime → Acetanilide

Ref: *Synth Commun, 2011, 41, 879-884.*

15) Microwave-assisted Pechmann reaction

The acid catalysed condensation of phenols with carboxylic acid or esters containing β-carbonyl functionality generates coumarins. This reaction is known as Pechmann reaction. Valizadeh et al (2009) reported the condensation of resorcinol and ethyl acetoacetate under microwave irradiation (140 W, 25 min) to generate 7-hydroxy-4-methylcoumarin (85%). This reaction utilized imidazolium dihydrogenphosphate as a catalyst.

HO, OH, O, O, H_3C, OC_2H_5, +, MW 140 W, 25 min, HO, O, O, CH_3

Resorcinol Ethyl acetoacetate 7-Hydroxy-4-methyl coumarin

Ref: *Phosphorous, Sulfur and Silicon, 2009, 184, 3075-3081.*

16) Microwave-assisted Knoevenagel condensation

The base (e.g. piperidine) catalysed condensation of active hydrogen containing compound with carbonyl compound is known as Knoevenagel condensation. In general, this reaction produces α, β-unsaturated ketone (conjucated enone). Gupta et al (2007) reported the microwave assisted synthesis of cinnamic acid through Knoevenagel condensation. The condensation of benzaldehyde and melonic acid in the presence of TBAB and potassium carbonate under microwave irradiation (900 W, 5 min) produced cinnamic acid (85%).

O, CH_3, +, COOH, COOH, MW 900 W, 5 min, CH_3, COOH

Acetophenone Melonic acid Cinnamic acid

Ref: *Arkivoc, 2007, 94-98.*

17) Microwave-assisted Baylis-Hilmann reaction (Morita-Bayilis-Hilman reaction)

The coupling reaction between an aldehyde (electrophile) and and an activated alkene containg electron withdrawing group in the presence of Lewis base catalyst generates α-hydroxylalkylated compound. The coupling reaction is known as Baylis-Hilmann reaction. 1.4-diazabicyclo [2.2.2]octane (DABCO) is the most commonly useful catalyst. De Souza et al (2008) reported the microwave assisted synthesis of methyl 3-hydroxy-2-methylene-3-(4-nitrophenyl-propanoate through Baylis-Hilmann reaction. The DABCO catalysed reaction between 4-nitrobenzaldehyde and methylacrylate under microwave irradiation (60 min) produced methyl methyl 2-[hydroxy(4-nitrophenyl)methyl]prop-2-enoate (90%).

CHO
O_2N
4-Nitrobenzaldehyde
+ H_2C OCH_3 O
Methyl acrylate
MWI, 60 min
OH O OCH_3 CH_2 O_2N
Methyl 2-[hydroxy(4-nitrophenyl) methyl]prop-2-enoate

Ref: *Letters in Organic Chemistry, 2008, 5, 379-382.*

18) Microwave-assisted Hantsch 1,4-dihydropyridine synthesis

The condensation of aryl aldehyde, β-ketoester and ammonia (amines) in acid catalysed reaction produces 1, 4-dihydropyridines. This reaction is known as Hantsch 1, 4-dihydropyridine synthesis. Maru et al (2019) reported the synthesis of 1, 4-dihydropyridine. The microwave assisted (400 W, 15 min) catalyst free cyclo-condensation of 3-bromo-4-hydroxy-5-methoxybenzaldehyde, ethyl acetoacetate and ammonium carbonate to generated the 1, 4-dihydropyridine (97%).

3-Bromo-4-hydroxy-5-methoxybenzaldehyde + 2 Ethyl acetoacetate + $(NH_4)_2CO_3$ Ammonium carbonate → (MW 400 W, 15 min) Diethyl-2,6-dimethyl-(3-bromo-4-hydroxy-5-mewthoxy)-4-phenyl-1,4-dihydropyridine-3,5-dicarboxylate

Ref: *Chemistry Select, 2019, 4, 774-782.*

19) Microwave-assisted Ullman reaction

The copper metal catalysed coupling reaction between two molecules of aryl halides to form biaryl compound is known as Ullman reaction. Gadda et al (2012) reported the microwave assisted synthesis of biphenyl compound. Gadda et al (2012) reported the microwave assisted synthesis biphenyl compound. The coupling reaction between two molecules of iodobenzene using Pd/C and sodium hydroxide base under microwave irradiation (150 °C, 30 min) generated biphenyl compound (97%).

Iodobenzene → (150 °C, 30 min; Pd/C, NaOH) Biphenyl

Ref: *Synth Commun, 2012, 42, 1259-1267.*

20) Microwave-assisted Williamsons ether synthesis

The reaction between alkoxide ion (nucleophile) and alkyl halide and generates ether through halide ion displacement (S_N2 reaction). This reaction is known as Williamsons ether synthesis. Baar et al (2010) reported the microwave assisted synthesis of 2-ethoxynaphthalene. The reaction between potassium 2-naphtholate and ethyl iodide in methanol under microwave irradiation (130 °C, 40s) produced 2-ethoxynaphthalene (47%).

2-Naphthol + $H_3C{-}CH_2{-}I$ (Ethyliodide) $\xrightarrow{130\ °C,\ 40s}$ 2-Ethoxynaphthalene (OC_2H_5)

Ref: *J Chem Edu, 2010,87(1), 84-86*

21) Microwave-assisted Paal-Knorr pyrrole synthesis

The generation of pyrrole derivatives from 1,4-carbonyl compounds and ammona (or primary amines) in acid catalysed reaction is known as Paal-Knorr pyrrole synthesis. Aghpoor et al (2018) reported the calcium chloride catalysed microwave assisted synthesis of *N*-substituted pyrroles. The reaction between 2, 5-hexanone and 4-bromoaniline under microwave irradiation (420 W, 10 min) produced 1-(4-bromophenyl)-2,4-dimethyl-pyrrole (100%).

2,5-Hexanone + 4-Bromoaniline $\xrightarrow{MW\ 420,\ 10\ min}$ 1-(4-Bromophenyl)-2,5-dimethtyl-1*H*-pyrrole

Ref: *Research on Chemical Intermediates, 2018, 44(7), 4063-4072*

22) Microwave-assisted Arbuzov reaction (Michaelis-Arbuzov reaction)

The reaction between a trialkylphosphite (trivalent) and an alkyl halide produces a phosphonate (pentavalent). This reaction is known as Arbuzov reaction. Heglivich et al (2012) reported the nickel chloride catalysed synthesis of diethyl phenylphosphonate (90%) by the reaction between the bromobenzene and triethylphosphite under microwave irradiation (160 ºC, 2 h).

Br + $(C_2H_5O)_3P$ —MW 160 °C, 2 h→ $O{=}P(OC_2H_5)_2$

Iodobenzene Triethyl phosphite Diethyl phenylphosphonate

Ref: *HeteroatomChemistry, 2012, 23(6), 574-582*

Important microwave assisted drug synthesis

1) Microwave-assisted synthesis of cephalosporin derivatives

Cephalosporins are broad spectrum antibacterial gents which produces low toxicity. Kidwai et al (2000) reported the synthesis of 7-aminosubstituted cephalosporinic acid derivatives through microwave assisted reaction. The reaction between 7-amino cephalosporanic acid and 4-pyridine carboxylic acid under microwave irradiation (120 s) produced of 7-pyridine formylamino cephalosporinic acid (92%).

COOH + H_2N, S, N, O, COOH, O, OCH_3 —MWI 120 s→ N, H, N, O, S, N, O, COOH, O, OCH_3

Pyridine-4-carboxylic acid 7-Amino cephalosporanic acid 7-Pyridine formylamino cephalosporanic acid

Ref: *Monatshefte fuEr Chemie, 2000, 131, 937-943*

2) Microwave-assisted synthesis of cisplatin

Cisplatin (cisplatinum) is chemically *cis*-diamminedichlroplatinum. It is used in the treatment of baldder, lung, ovarian and testicular cancers. Petruzzella et al (2015) reported the microwave assisted synthesis for the cisplatin. The reaction between potassium tetrachloroplatinate, potassium chloride and ammonium acetate under microwave irradiation (100 °C, 14 min) produced cisplatin (47%).

$2K^+ [PtCl_4]^{2-}$ (Potassium tetrachloroplatinate) → KCl, CH_3COONH_4, MWI 100 °C, 14 min → Cisplatin

Ref: *Dalton Trans, 2015, 44, 3384-3392*

3) Microwave-assisted synthesis of phenytoin

Phenytoin is chemicall 5,5,-diphenylhydantoin. It is useful in the treatment of wide variety of seizure disorders. Gbaguidi et al (2011) reported the microwave assisted synthesis of phenytoin. The mixture of benzil, urea and potassium hydroxide in dimethylsulphoxide upon microwave irradiation (1100 W, 1.5 min) produced phenytoin (87%).

Benzil + Urea → MW 1100 W, 1.5 min → Phenytoin

Ref: *African J Pure App Chem, 2011, 5(7), 168-175*

4) Microwave-assisted synthesis of procaine

Procaine is chemically an ester of 4-aminobenzoic acid. It is useful in inducing local anaesthesia. Noditi et al (2014) reported the synthesis of procaine through microwave irradiation. The reaction between ethyl 4-aminobenzoate (benzocaine) in sodium ethoxide and 2-(diethylamino)ethanol upon microwave heating (700 W, 12 min) produced procaine (86%).

Ethyl 4-aminobenzoate (Benzocaine) + Diethylamino ethanol → Procaine (MW 700 W, 12 min)

Ref: *Rev Chim, 2014, 65(1), 64-67*

5) Microwave-assisted synthesis of aspirin

Aspirin is chemically acetylsalicylic acid. It is used as an analgesic, antipyretic, antiplatelet and anti-inflammatory agent. Montes et al (2006) reported the solvent free synthesis of aspirin using microwave heating. The acetylation reaction between the salicylic acid and acetic anhydride upon microwave irradiation (80% MWI, 13 min) offered aspirin (97%).

Salicylic acid + $(CH_3CO)_2O$ (Acetic anhydride) → Aspirin (MW 80%, 13 min)

Ref: *J Chem Edu, 2006,83(4), 628-631*

Optimal Conditions for Microwave-Assisted Organic Synthesis

Microwave-assisted reactions, in comparison with prolonged thermal conventional heatings are much-faster (requires fewer minutes) and yields high quality products.

Similar to conventional organic reactions, the microwave reactions also demands the reaction optimization. The parameters namely solvent, power, temperature, reaction time, etc need to be optimized for the efficient reactions. The parameters also include the reactant stoichiometry, and catalyst concentration. In order to utilize the full benefits of microwave heating, the following sections need to be done and conditions need to be maintained.

1) **Solvent**: The solvents used in microwave reactions should possess dielectric heating property. Dielectric heating ensures the conversion of electromagnetic energy into efficient heating. The ability of the solvent dielectric property is indicated by tan δ. The solvents with high tan δ provide rapid heating. However the solvent with low tan δ also used to some extent in microwave heating depending on necessity of reaction conditions can be utilized (Table 2).

Table 2 Microwave solvents with dielectric constant values

High tan δ (>0.5)		**Medium tan δ (0.1- 0.5)**		**Low tan δ (<0.1)**	
Ethylene glycol	1.35	2-Butanol	0.45	Chloroform	0.09
Ethanol	0.94	Dichlorobenzene	0.28	Acetonitrile	0.06
DMSO	0.83	N-Methyl-2-pyrollidine	0.28	Ethylacetate	0.06
2-Propanol	0.80	Acetic acid	0.17	Acetone	0.05
Formic acid	0.72	DMF	0.16	THF	0.05
Methanol	0.70	Dichloroethane	0.13	DCM	0.04
Nitrobenzene	0.59	Water	0.12	Toluene	0.04
1-Butanol	0.57	Chlorobenzene	0.10	Hexane	0.02

- Polar solvents (e.g. methanol, ethanol) are very suitable for microwave reactions (heating). The dielectric properties of ionic liquids make them highly suitable for use as solvents or additives in microwave-assisted organic synthesis.
- Non-polar solvents such as toluene, dioxane, THF also can be used, if other components in the reaction mixture respond to microwave energy.
- Low boiling point solvents (e.g. methanol, dichloromethane and acetone) develops pressure in the vessel, hence offers lower reaction temperatures.

2) **Reaction phase:** Microwave reactions can be performed in solution, solid, solvent free (dry) and polymer supported medium. Microwave reactions are beneficial under solvent free conditions. This dry reaction (neat) offers several benefits. Mainly it reduces the explosive risks associated with the solution phase reactions. This kind of reaction conditions offer the microwave induction for the reactants and reagents, hence promotes the rate of reaction (efficient).

   ***Inert atmosphere*:** The reactions with sensitivity for oxygen and moisture requires inert gas environment. Purifies argon and liquid nitrogen are the most common inert gases used in the organic reactions. But, microwave reactions do not require such conditions. The microwave reaction vessel can be flushed (washed) with the inert gas, if it is necessary.

3) **Power (temperature):** Microwave power has direct influence on the reaction progress and rate of reaction. The exothermic nature of the reaction decides the power requirement. Hence, chemical reactions like dehydration require high power. In general, different instruments use different magnetron and their capacity slightly differes. This in turn decides the power levels to be set for given reaction. The most general power set up is given in the Table 3.

Table 3 Microwave power levels

Power	Temperature	
	in degree Celsius (°C)	**in Watts**
90 - 100	220 - 260	900 (100%)
80	190 - 220	600 (67%)
70	175 -190	
50 - 60	150 - 175	450 (50%)
30 - 40	110 - 150	300 (33%)
20	90 -110	180 (20%)
10	65 -90	100 (11%)

Microwave heating through the super heating property enhances the reactivity and product purity (yield). Super heating refers to the ability of solvents to boil above their boiling points. Microwaves also avoid the surface heating and, over heating of the specific components. In other words, microwave offers the uniform heating to the mixture causes molecular friction and facilitates the reaction rates.

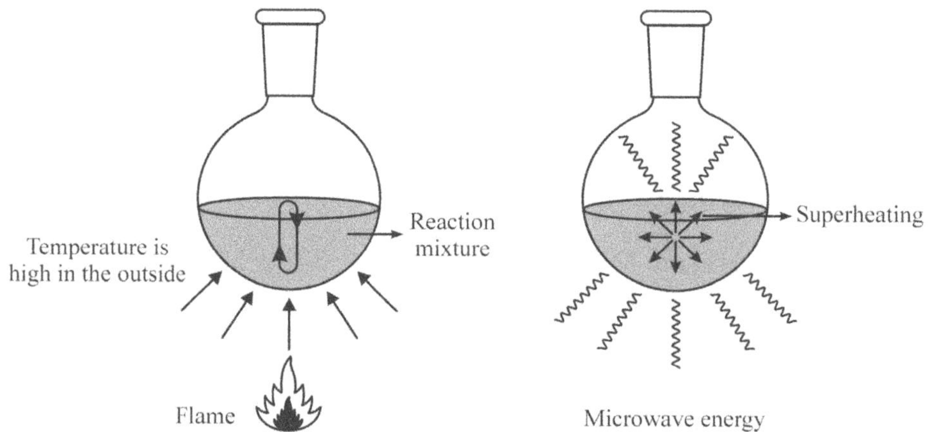

***Intermittent cooling*:** This is very important for the exothermic reactions. Intermittent cooling prevents the decomposition of reactants. In general, the low temperature reactions do not require intermittent cooling.

In microwave heating, the heat is generated inside the material. There is no direct contact between the energy source and the sample undergoing heating. This offers higher heating rate. In conventional heating, the outside of the vessel is heated first and it leads to thermal stress and cracking in the glassware.

4) **Reaction vessel**: The reaction vessels should be selected based on the quantities of the reactants (in grams, moles). Specially designed glassware are available for the microwave reactions. The design aspect also includes the pressure withstanding property (more than 30 bars). In general the capacity of the reaction vessel should be less than 50 ml. It can still lower depending upon the reactant concentration. Microwave transparent materials such as Teflon, polystyrene, pyrex or borosilicate glass are useful in the vessel fabrication. These materials absorb the radiation poorly and with stand at higher temperatures.

5) **Stirring:** To stir well is the universal instruction for chemical reactions. It is very essential for heterogeneous and solid phase reactions. Stirring the reaction mixture with the aid of in-build stirring enhances the heating process. In case of solvent less reactions (solid-phase), the mixture can be stirred using spatula during intermittent cooling

6) **Reaction time:** Microwave-assisted organic reactions are much-faster. Hence, most of the reactions can be completed in minutes (<15 minutes) and some of the reaction occur in seconds. The duration of the microwave heating can be optimized depending upon the reaction nature and instrument (model). The property of super heating in microwaves offers the multi fold increase in the reaction rate.

7) **Stoichiometry:** It refers to the quantitative relationship between the reactant and products. The relative quantities of the reactants influence the chemical

reactions. The quantities determine the equilibrium of the reactions. This is also applicable for microwave assisted reactions.

- Unimolecular reactions are independent of reactant concentration.
- Bi and trimolecular reactions depend on the reactant concentration. Higher concentrations enable the reaction rate.

Special Guidelines

a) General

- Special precautions are required for handling the reaction flasks.
    - Monitor through the glass window, alone.
    - Never remove the reaction vessel immediately after the microwave irradiation. Allow it to cool for few seconds.
- Monitor the progress of reactions through thin-layer chromatography (TLC)

b) To promote the reaction/drive the reaction to completion:

The difficult reactions can be facilitated by adopting the following:

- Increase the reaction duration
- Increase the reaction temperature (Watts)
- Increase the concentration of reactants and reagents
- Replacing the solvent (increasing the polarity, dipolar constants)
- Replacing the reagents with more microwave absorbing reagents

c) To prevent the decomposition of reactants and products

The difficult reactions can be facilitated by adopting the following:

- Lower the temperature (Waats)
- Shorten the reaction time
- Decrease the concentration of reactants and reagents
- Replacing the reagents with less microwave absorbing reagents

Microwave Instruments

a) Domestic microwave ovens

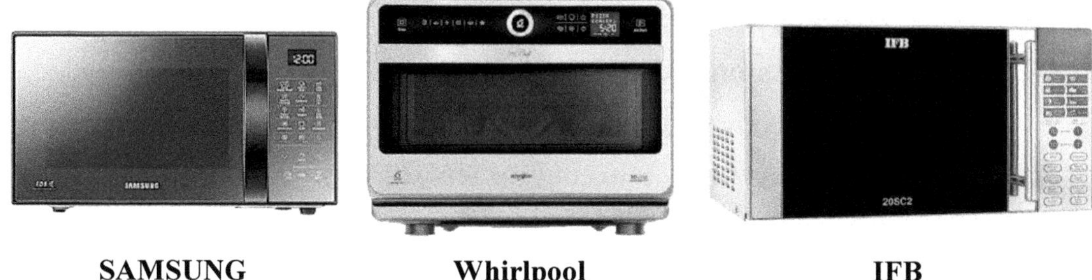

SAMSUNG Whirlpool IFB

b) Microwave synthesizers (specially designed for chemical synthesis)

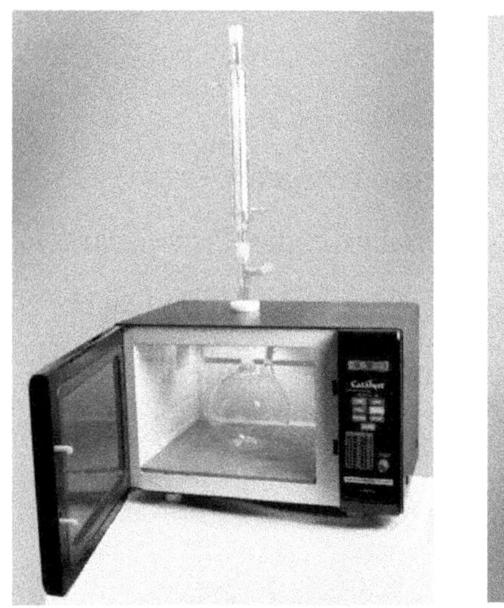

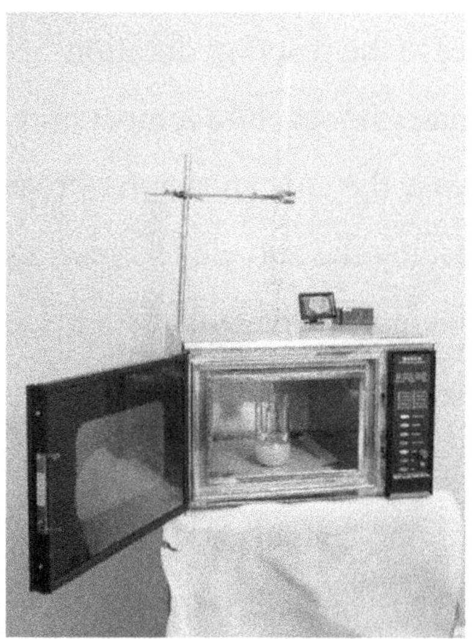

PART - 2

ORGANIC SYNTHESIS

MICROWAVE INTENSITY AND POWER (WATTS)

In general, the microwave intensity and the corresponding power in Watts used in the optimization of reactions is as depicted in the below table. However, this may change depending upon make of the instrument. For instance, domestic microwave wave instruments with 600 to 1200 W are also available. In fact, higher Wattage instruments are costlier than the low Wattage instruments.

Hence, the user needs to realign with the instrument of their choice for the better results. In case, instruments with the Intensity and power is as mentioned below, the reactions proceeds very well with the reaction settings mentioned in this practical manual.

MW % Intensity	***Power (Watts)***
10	85
20	170
30	255
40	340
50	425
60	510
70	595
80	680
90	765
100	850

The **Watt** (symbol: **W**) is a unit of power. In the International System of Units (SI) it is defined as a derived unit of 1 joule per second, and is used to quantify the rate of energy transfer.

ACETYLATION

Acetylation involves the replacement of active hydrogen of hydroxyl and amino (1° & 2°) groups by an acetyl group of acetylating agent. Acetylation is also known as ethanoylation. Alcohols (R-OH), phenols (Ar-OH), primary amines ($R-NH_2$) and secondary amines (R-NH-R) can be acetylated. Alcohol and phenol undergo O-acetylation. Amines (1° & 2°) form N-acetylation products.

EXPERIMENT 1

Preparation of Aspirin

Principle: Acetylation of salicyclic acid with acetic anhydride in presence of glacial acetic acid generates aspirin (acetyl salicylic acid). Sulphuric acid and phosphoric acid can be used as an alternative for glacial acetic acid. Acetyl group of acetic anhydride replaces the active hydrogen of hydroxyl group. This reaction is typical example for O-acetylation. One drop of phosphoric acid or magnesium sulphate (1%) can be added as a catalyst. These catalysts absorb microwave radiations. The product with high purity without recrystallization is the advantage of this method. Magnesium sulfate offers excellent crystals of aspirin.

$$\text{Salicylic acid (OH, COOH)} + (CH_3CO)_2O \xrightarrow[-CH_3COOH]{\text{MW 425 W, 4 min};\ H_2SO_4} \text{Aspirin (}OCOCH_3\text{, COOH)}$$

Salicylic acid Acetic anhydride Aspirin

Reaction Mechanism

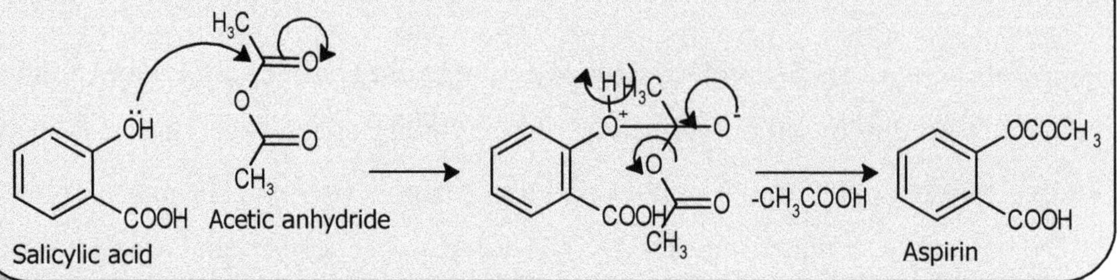

Chemicals required

- Salicylic acid
- Acetic anhydride
- Sulphuric acid

Procedure:

1) Load the mixture of salicylic acid (1 g, 7.24 mmol), acetic anhydride (2 ml) and sulphuric acid (0.2 ml) into microwave flask (beaker, 50 ml).
2) Place the flask into the Microwave oven and switch on the magnetic stirrer.
   - In case of Microwave synthesizer, attach the reflux condenser to the flask.
   - In case of domestic microwave oven, cover the flask with a petridish loaded with ice cubes (provides condensing effect), alternatively funnel can be used. Baker containing cold water can be used as a heating sink.
3) Irradiate with microwaves at 50% intensity for 4 minutes, with intermittent cooling for one minute after 3 minutes of microwave irradiations.
4) Turn off the power supply to the microwave oven.
5) After few minutes, remove the flask from the microwave oven.
6) Cool and add cold water (7.5 ml) in to the flask (removes excess of acetic anhydride).
7) Filer the separated product and recrystallize from 50% methanol.
8) The melting point of pure aspirin is 135-137 °C.

Comparison of conventional and microwave method

	Conventional	*Microwave*
Reaction time	15 min	4 min
Reaction temperature	50-60 °C	50% intensity (425 W)
Colour	White crystals	
Melting point	135-137 °C	
Percentage yield	96%	97%

Post Lab Questions:

1) What is acetylation?
2) What is the role of concentrated sulphuric acid in the preparation of aspirin?
3) Draw the structure of aspirin.

EXPERIMENT 2

Preparation of Acetanilide

Principle: Aniline upon acetylation with acetylating reagent such as acetic anhydride or acetyl chloride leads to the formation of acetanilide in presence of glacial acetic acid. Acetyl group of acetylating agent, acetylates the nitrogen of aniline. This is typical example for N-acetylation. However, this reaction can be driven without any catalyst.

$$C_6H_5-NH_2 + (CH_3CO)_2O \xrightarrow{\text{MW 510 W, 8 min}} C_6H_5-NH-\overset{O}{\overset{\|}{C}}-CH_3$$

Aniline | Acetic anhydride | Acetanilide

Reaction mechanism

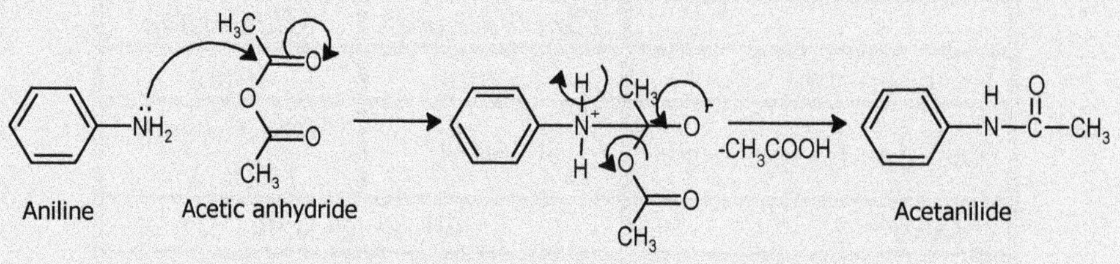

Aniline + Acetic anhydride → [intermediate] —($-CH_3COOH$)→ Acetanilide

Chemicals required

- Aniline
- Glacial acetic acid
- Acetic anhydride

Procedure:

1) Load the mixture of aniline (1.2 ml, 13 mmol) and acetic anhydride (3 ml) into a 50 ml flask. This reaction requires dry conditions.
2) Place the flask into the Microwave oven and switch on the magnetic stirrer.
    - In case of Microwave synthesizer, attach the reflux condenser to the flask.
    - In case of domestic microwave oven, cover the flask with a petridish loaded with ice cubes (provides condensing effect), alternatively funnel can be used. Baker containing cold water can be used as a heating sink.
3) Irradiate the reaction mixture at 60% microwave intensity for 8 minutes, with intermittent cooling for one minute after 3 minutes of microwave irradiations.
4) During cooling stir the reaction mixture with spatula.
5) Turn off the power supply to the microwave oven.
6) After few minutes, remove the flask from the microwave oven.
7) Add 20 ml of water and allow to stand for 6 minutes.
8) Filter the precipitated white solid, wash with cold water and dry.
9) Recrystallize the product with hot water.

Note: Monitor the progress of reaction through thin-layer chromatography (TLC).

Comparison of conventional and microwave method

	Conventional	*Microwave*
Reaction time	30 min	8 min
Reaction temperature	50-60 °C	60% intensity (510 W)
Colour	white crystals	
Melting point	114-115 °C	
Percentage yield	95%	96%

Safety precautions

- Avoid exposure to glacial acetic acid and acetic anhydride vapors.
- Collect the flask few minutes after opening the window of microwave oven.

Post Lab Questions:

1) What is N-alkylation?
2) Name the acetylating agent used in the preparation of acetanilide.

EXPERIMENT 3

Preparation of *N*-Acetylanisidine

Principle: Acetylation of anisidine by acetic anhydride-acetic acid generates N-acetylanisidine. The amine nitrogen of anisidine undergo acetylation (N-acetylation).

$$H_3C\text{-}O\text{-}C_6H_4\text{-}NH_2 + (CH_3CO)_2O \xrightarrow[CH_3COOH]{\text{MW 425 W, 5 min; MW 255 W, 2 min}} H_3C\text{-}O\text{-}C_6H_4\text{-}NH\text{-}C(=O)\text{-}CH_3$$

Anisidine Acetic anhydride N-Acetylanisidine

Reaction mechanism

Chemicals required

- Anisidine
- Acetic anhydride
- Acetic acid (80%)

Note: Monitor the progress of reaction through thin-layer chromatography (TLC).

Procedure:

1) Load the mixture of anisidine (1 g, 8.1 mmol), acetic anhydride (2.5 ml) and 10 ml of acetic acid (80%) into microwave flask (beaker, 50 ml).
2) Place the flask into the Microwave oven and switch on the magnetic stirrer.
    - In case of Microwave synthesizer, attach the reflux condenser to the flask.
    - In case of domestic microwave oven, cover the flask with a petridish loaded with ice cubes (provides condensing effect), alternatively funnel can be used. Baker containing cold water can be used as a heating sink.
3) Irradiate the mixture with microwaves at 425 W intensity for 5 minutes, with intermittent cooling for one minute after 3 minutes of microwave irradiations.
4) During cooling stir the reaction mixture with spatula.
5) Cool and irradiate once again at 255 W for 2 minutes to complete the reaction.
6) Turn off the power supply to the microwave oven.
7) After few minutes, remove the flask from the microwave oven.
8) Add 20 ml of water and allow it to stand for 5 minutes.
9) Filter the precipitated white solid, wash with cold water and dry.
10) Recrystallize the product with chloroform or hexane.

Comparison of conventional and microwave method

	Conventional	***Microwave***
Reaction time	30 min	7 min
Reaction temperature	50-60 °C	425 W (5 min) 255W (2 min)
Colour	White crystals	
Melting point	128-130 °C	
Percentage yield	94%	96%

Safety precautions

- Avoid exposure to glacial acetic acid and acetic anhydride vapors.
- Collect the flask few minutes after opening the window of microwave oven.

Post Lab Questions:

1) What is N-alkyltion?
2) What is the role of glacial acetic acid in this preparation?

EXPERIMENT 4

Preparation of Paracetamol

Principle: The acetylation of *p*-aminophenol with acetic anhydride in the presence of glacial acetic acid forms paracetamol (N-acetylation product, acetaminophen). N-Acetylation is preferred over the acetylation on hydroxyl group under microwave irradiation conditions.

$HO-C_6H_4-NH_2 + (CH_3CO)_2O \xrightarrow{340\ W,\ 5\ min} HO-C_6H_4-NH-CO-CH_3 + H_3C-CO-O-C_6H_4-NH-CO-CH_3$

p-Aminophenol, Acetic anhydride → Paracetamol (major product) + Minor product

Reaction mechanism

p-Amino phenol + Acetic anhydride → (tetrahedral intermediate) → $-CH_3COOH$ → Paracetamol

Chemicals required

- *p*-Aminophenol
- Acetic anhydride
- Acetic acid (80%)

Procedure:

1) Load the mixture of *p*-aminophenol (1 g, 9.16 mmol) and acetic anhydride (2 ml) in acetic acid (80%, 10 ml) into microwave flask (beaker, 50 ml).
2) Place the flask into the Microwave oven and switch on the magnetic stirrer.
   - In case of Microwave synthesizer, attach the reflux condenser to the flask.
   - In case of domestic microwave oven, cover the flask with a petridish loaded with ice cubes (provides condensing effect), alternatively funnel can be used. Baker containing cold water can be used as a heating sink.
3) Irradiate the mixture with microwaves at 40% intensity for 5 minutes, with intermittent cooling for one minute after 3 minutes of microwave irradiations.
4) Turn off the power supply to the microwave oven.

5) After few minutes, remove the flask from the microwave oven.
6) Add 20 ml of water and allow to stand for 5 minutes.
7) Filter the precipitated white solid, wash with cold water and dry.
8) Recrystallize the product with chloroform to get pure paracetamol.

Note: Monitor the progress of reaction through thin-layer chromatography (TLC).

Comparison of conventional and microwave method

	Conventional	*Microwave*
Reaction time	1 h	5 min
Reaction temperature	50-60 °C	40% intensity (340 W)
Colour	White	
Melting point	169-170 °C	
Percentage yield	88%	90%

Post Lab Questions:

1) What is the therapeutic use of paracetamol?
2) What is the role of glacial acetic acid in this preparation?

EXPERIMENT 5

Preparation of α-Glucose Pentaacetate

Principle: Acetylation of D-glucose generates glucose pentaacetate. Glucose exists in the form of α and β - cyclic anomers, differing in the arrangement of -H and -OH on the anomeric carbon. Both the form undergo aceylation. Acetylation of glucose is one of the simplest methods to identify the two forms of glucose. The catalysts zinc chloride (acidic) and sodium acetate (basic) influences the stereochemistry of the reaction. Zinc chloride produces α-acetate as the main product. The basic sodium acetate produce β-acetate as the major product.

D-Dlucose + $(CH_3CO)_2O$ Acetic anhydride —(MW 510 W, 15 min, $ZnCl_2$)→ α-Glucose penataacetate

Chemicals required

- D-Glucose
- Acetic anhydride
- Zinc chloride (anhydrous)

Procedure:

1) Load the mixture of powdered zinc chloride (0.25 g, 1.8 mmol), acetic anhydride (5 ml) and glucose (1 g, 5.5 mmol) into microwave flask (beaker, 50 ml) and shake thoroughly.
2) Place the flask into the Microwave oven and switch on the magnetic stirrer.
   - In case of Microwave synthesizer, attach the reflux condenser to the flask.
   - In case of domestic microwave oven, cover the flask with a petridish loaded with ice cubes (provides condensing effect), alternatively funnel can be used. Baker containing cold water can be used as a heating sink.
3) Irradiate with microwaves at 60% intensity for 15 minutes, with intermittent cooling for one minute after 3 minutes of microwave irradiations.
4) During cooling stir the reaction mixture with spatula.
5) Turn off the power supply to the microwave oven.

6) After few minutes, remove the flask from the microwave oven.
7) Add cold water (7.5 ml) in to the reaction mixture with stirring.
8) Continue stirring until the oily liquid solidifies.
9) Filter the separated product, wash with cold water.
10) Purify the product by recrystallization with methanol (50%).

Note: Monitor the progress of reaction through thin-layer chromatography (TLC).

Comparison of conventional and microwave method

	Conventional	***Microwave***
Reaction time	1 hr	15 min
Reaction temperature	50-60 °C	60% intensity (510 W)
Colour	White	
Melting point	112-114 °C	
Percentage yield	76%	84%

Post Lab Questions:

1) How else might one make glucose pentaacetate?
2) What advantages or disadvantages of other procedures as compared to the acetic anhydride method?

ESTERIFICATION

Esters are one of the most important class of organic compounds. A large number of esters occur in flowers and fruits with different fragrance. They are used in many synthetic products such as perfumes, pesticides, fibers, solvents, drugs, pharmaceuticals and plasticizers.

EXPERIMENT 6

Preparation of Benzocaine

Benzocaine history: The Andean society (south america) have been using the leaves of the 'Coco bush' (*Erythroxylon coca*) as a mild stimulant. Chemists isolated the alkaloid "cocaine" from the cocoa leaves. This alkaloid was used as an anaesthetic in surgery and dentistry. Cocaine causes irreversible damage to the central nervous system (CNS). Cocaine analogues were prepared (after the structure of cocaine was established in 1910) to lower the CNS toxicity. The ethyl ester of *p*-aminobenzoic acid, named as benzocaine was good example for cocaine analogue.

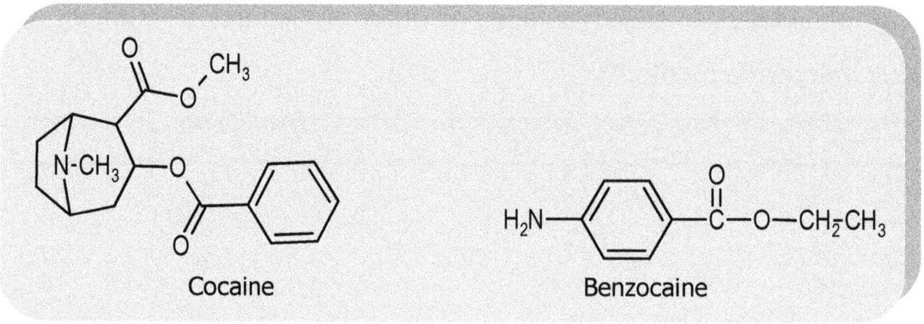

Cocaine

Benzocaine

Principle: Chemically benzocaine is an ethyl ester of *p*-amino benzoic acid. The preparation of Benzocaine involves esterification reaction between the *p*-amino benzoic acid and dry ethanol in presence of mineral acid catalyst (sulphuric acid or dry hydrogen chloride). The benzocaine can be prepared using microwave heating. It requires microwave energy (50%) for 6 minutes.

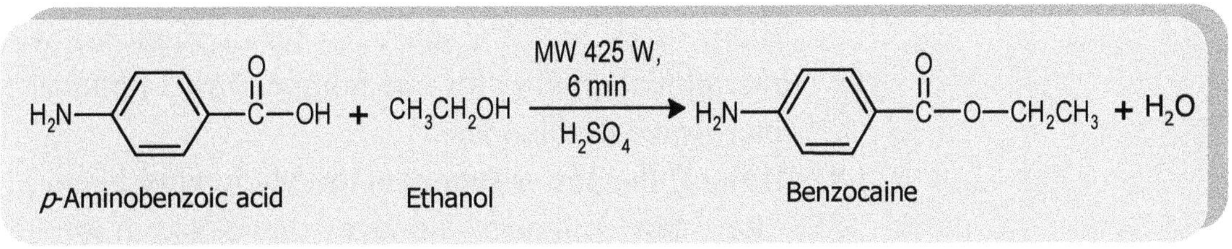

$H_2N-C_6H_4-COOH + CH_3CH_2OH \xrightarrow[H_2SO_4]{MW\ 425\ W,\ 6\ min} H_2N-C_6H_4-CO-O-CH_2CH_3 + H_2O$

p-Aminobenzoic acid   Ethanol   Benzocaine

Reaction mechanism

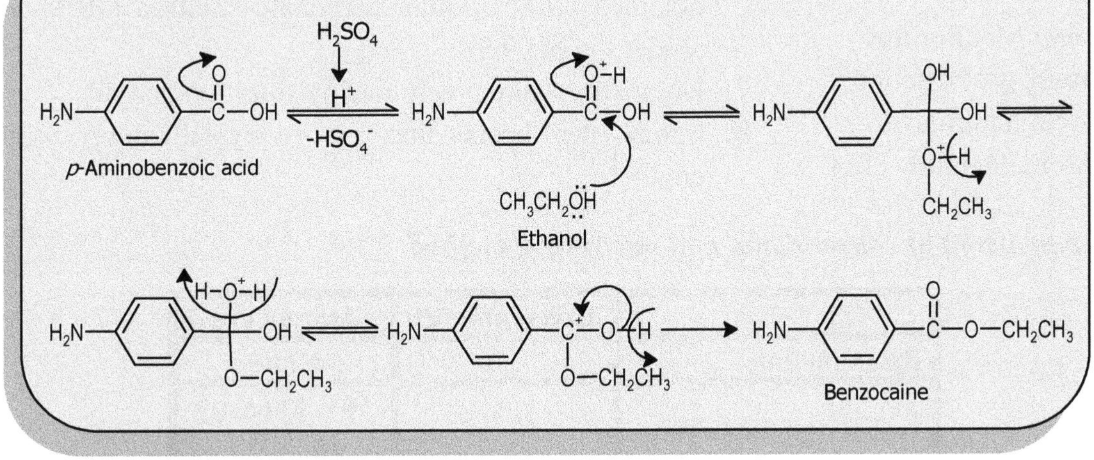

H_2SO_4, H^+, $-HSO_4^-$

p-Aminobenzoic acid

$CH_3CH_2\ddot{O}H$

Ethanol

Benzocaine

Chemicals required

- *p*-Aminobenzoic acid
- Absolute ethanol
- Concentrated H_2SO_4
- Sodium carbonate (10%) solution

Procedure:

1) Load *p*-aminobenzoic acid (0.5 g, 3.6 mmol), dry ethanol (6 ml) and concentrated sulphuric acid (1 ml) into microwave flask (beaker, 50 ml) and mix well.
2) Place the flask into the Microwave oven and switch on the magnetic stirrer.
   - In case of Microwave synthesizer, attach the reflux condenser to the flask.
   - In case of domestic microwave oven, cover the flask with a petridish loaded with ice cubes (provides condensing effect), alternatively funnel can be used.

Beaker containing cold water can be used as a heating sink.

3) Irradiate the reaction mixture present in the flask with Microwaves at 50% intensity for 6 minutes, with intermittent cooling for one minute after 3 minutes of microwave irradiations.
4) Turn off the power supply to the Microwave oven.
5) After few minutes, remove the flask from the Microwave oven.
6) Cool and alkalify the pale brown colored liquid obtained with sodium carbonate solution (10%) to render the pH 9.0.
7) Isolate the white precipitate by filtration and dry.
8) Purify the benzocaine by re-crystallization with ethanol.

Note: Monitor the progress of reaction through thin-layer chromatography (TLC).

Comparison of conventional and microwave method

	Conventional	*Microwave*
Reaction time	2 h	6 min
Reaction temperature	78 °C	50% intensity (425 W)
Colour	White	
Melting point	88-89 °C	
Percentage yield	76%	90%

Post Lab Questions:

1) What is the role of mineral acid in this reaction?
2) What are the other methods for the preparation of esters?
3) Mention the advantages and disadvantages of other esterification methods.
4) Why dry ethanol is preferred in this reaction?
5) Why alkalification is required in the benzocaine preparation.

EXPERIMENT 7

Preparation of Butamben

Principle: Butamben is an amino acid ester. The condensation of carboxylic acid group of *p*-aminobenzoic acid with alcoholic group of dry *n*-butanol in presence of sulphuric acid (catalyst) forms butamben. It has surface anesthetic property on skin and mucous membranes.

$$H_2N-C_6H_4-C(=O)-OH + C_4H_9OH \xrightarrow[H_2SO_4]{\text{MW 425 W, 8 min}} H_2N-C_6H_4-C(=O)-O-C_4H_9 + H_2O$$

p-Aminobenzoic acid   *n*-Butanol   Butamben

Chemicals required

- *p*-Aminobenzoic acid
- *n*-Butanol
- Concentrated sulphuric acid
- Sodium carbonate (10%) solution
- Diethyl ether

Procedure:

1) Load *p*-aminobenzoic acid (0.5 g, 3.6 mmol), *n*-butanol (3 ml) and concentrated sulphuric acid (1 ml) into microwave flask (beaker, 50 ml) and mix well.
2) Place the flask into the Microwave oven and switch on the magnetic stirrer.
   - In case of Microwave synthesizer, attach the reflux condenser to the flask.
   - In case of domestic microwave oven, cover the flask with a petridish loaded with ice cubes (provides condensing effect), alternatively funnel can be used. Beaker containing cold water can be used as a heating sink.
3) Irradiate the reaction mixture present in the flask with Microwaves at 50% intensity for 8 minutes, with intermittent cooling for one minute after 3 minutes of microwave irradiations.

4) Turn off the power supply to the Microwave oven.
5) After few minutes, remove the flask from the Microwave oven.
6) Cool and alkalify the pale brown colored liquid obtained with sodium carbonate solution (10%) to render the pH 9.0.
7) Extract the liquid with diethyl ether (2 X 20 ml) and evaporate the combed extracts.
   (*upon complete evaporation, a oily layer will appear*)
8) Cool the oily layer to obtain the creamy white precipitate.
9) Isolate the white precipitate by filtration and dry.
10) Purify the butamben by re-crystallization with ethanol.

Note: Monitor the progress of reaction through thin-layer chromatography (TLC).

Comparison of conventional and microwave method

	Conventional	***Microwave***
Reaction time	2 h	8 min
Reaction temperature	95 °C	50% intensity (425 W)
Colour	White	
Melting point	58-60 °C	
Percentage yield	76%	80%

Post Lab Questions:

1) What is the role of concentrated sulphuric acid in the butamben preparation?
2) Mention the role of diethyl ether in this preparation?
3) How the unreacted carboxylic acid and alcohol are removed in the esterification?

ELECTROPHILIC AROMATIC SUBSTITUTION

The substitution of a hydrogen of an aromatic ring by an electrophile (E^+) is known as electrophilic aromatic substitution (AES) reaction. Electrophile is also known as lewis acid. The electrophilic aromatic substitution reaction can be depicted as follows.

$$C_6H_5\text{-}H + E\text{-}Y \longrightarrow C_6H_5\text{-}E + H\text{-}Y$$

Benzene — Electrophile substituted benzene

Most important AES reactions are listed below.

- Nitration.
- Halogenation
- Sulphonation
- Friedel-Crafts alkylation.
- Friedel-Crafts acylation.

General reaction mechanism

Aroamtic electrophilic substitution reaction can be explained through the following three steps.

Step 1: Electrophile generation

$$E\text{-}Y \longrightarrow E^+ + Y^-$$

Electrophile

Step 2: Aromatic ring π-electrons attack on the electrophile

Aromatic system + E^+ (Electrophile) ⟶ arenium ion bearing H and E

Step 3: Formation of electrophilic substituted compound

$-H^+$

Electrophile substituted aromatic system

EXPERIMENT 8

Preparation of Nitrobenzene

Principle: Nitrobenzene is also called as oil of mirbane. It can be prepared by nitration of benzene, which involves aromatic electrophilic substitution reaction. Mixed acid reagent (mixture of concentrated nitric acid and sulphuric acid) generates nitronium ion (NO_2^+, electrophile). Concentrated sulphuric acid catalyses the formation of nitronium ion. Nitronium ion substitutes the hydrogen of the benzene and forms nitrobenzene.

Benzene + HNO_3 —(MW 425 W, 10 min, H_2SO_4)→ Nitrobenzene (NO_2)

Reaction mechanism

Generation of the electrophile:

$HO_3S{-}\ddot{O}{-}H + H\ddot{O}{-}N^+(O^-){=}O \longrightarrow O{=}N^+{=}O + {}^-OSO_3H + H_2O$

Sulphuric acid; Nitric acid; Nitronium ion

Electrophilic attack and deprotonation of carbocation:

Benzene + $O{=}N^{+}{=}O$ (Nitronium ion) → [carbocation intermediate, H, NO_2] $\xrightarrow{-H_2SO_4}$ Nitrobenzene (NO_2)

Chemicals required

- Benzene
- Concentrated nitric acid
- Concentrated sulphuric acid
- Sodium carbonate (10%) solution
- Diethyl ether

Procedure:

1) Load benzene (1 ml, 11.2 mmol), concentrated nitric acid (3.5 ml, 84 mmol) in a round bottom flask and add concentrated sulphuric acid (4.5 ml) into microwave flask (beaker, 50 ml).
2) Place the flask into the Microwave oven and switch on the magnetic stirrer.
   - In case of Microwave synthesizer, attach the reflux condenser to the flask.
   - In case of domestic microwave oven, cover the flask with a petridish loaded with ice cubes (provides condensing effect), alternatively funnel can be used. Beaker containing cold water can be used as a heating sink.
3) Irradiate the reaction mixture present in reaction flask with Microwaves at 50% intensity for 10 minutes, with intermittent cooling for one minute after 3 minutes of microwave irradiations.
4) Turn off the power supply to the Microwave oven.
5) After few minutes, remove the flask from the Microwave oven.
6) Extract the liquid with diethyl ether (2 X 20 ml) and evaporate the combined extracts.
7) Cool the oily layer to obtain the yellow oil of nitrobenzene.

Note: Monitor the progress of reaction through thin-layer chromatography (TLC).

Comparison of conventional and microwave method

	Conventional	*Microwave*
Reaction time	2.5 h	10 min
Reaction temperature	50-60 °C	50% intensity (425 W)
Colour	Pale yellow oil	
Melting point	211 °C	
Percentage yield	78%	86%

Post Lab Questions:

1) What is nitration?
2) What are the components of nitrating mixtures?
3) Why aromatic nitro compounds are yellow in colour?

EXPERIMENT 9

Preparation of *m-di*-Nitrobenzene

Principle: Nitrobenzene on nitration (AES) gives *m-di*-nitrobenzene. It involves the replacement of hydrogen atom on aromatic ring by nitronium ion (electrophile). The nitro group present in nitro benzene isc meta orienting, hence permits AES at metaposition and forms *m-di*-nitrobenzene.

NO_2 (Nitrobenzene) + HNO_3 $\xrightarrow[H_2SO_4]{\text{MW 425 W, 10 min}}$ NO_2, NO_2 (*m*-Dinitrobenzene)

Nitrobenzene

m-Dinitrobenzene

Chemicals required

- Nitrobenzene
- Concentrated sulphuric acid
- Fuming Nitric acid

Procedure:

1) Load nitrobenzene (1.25 ml, 14 mmol), concentrated nitric acid (3.5 ml) in a round bottom flask and add concentrated sulphuric acid (4.5 ml) into microwave flask (beaker, 50 ml).
2) Place the flask into the Microwave oven and switch on the magnetic stirrer.
   - In case of Microwave synthesizer, attach the reflux condenser to the flask.
   - In case of domestic microwave oven, cover the flask with a petridish loaded with ice cubes (provides condensing effect), alternatively funnel can be used. Beaker containing cold water can be used as a heating sink.
3) Irradiate the reaction mixture present in reaction flask with Microwaves at 50% intensity for 10 minutes, with intermittent cooling for one minute after 3 minutes of microwave irradiations.
4) Turn off the power supply to the Microwave oven.
5) After few minutes, remove the flask from the Microwave oven.
6) Cool the reaction mixtures and pour into beaker containing cold water.
7) Wash the precipitate with water and dry.
8) Purify the isolated product by re-crystallization with ethanol.

Note: Monitor the progress of reaction through thin-layer chromatography (TLC).

Comparison of conventional and microwave method

	Conventional	*Microwave*
Reaction time	1 h	10 min
Reaction temperature	50-60 °C	50% intensity (425 W)
Colour	Pale yellow	
Melting point	89-90 °C	
Percentage yield	76%	85%

Post Lab Questions:

1) What is AES?
2) Explain the general reaction mechanism involved in the AES.
3) Outline how nitronium ion (electrophile) is generated?

EXPERIMENT 10

Preparation of *p*-Bromoacetanilide

Principle: Acetanilide upon bromination (AES) produces *p*-bromoacetanilide. Bromine in glacial acetic acid produces the electrophile, Br^+. Bromine electrophile replaces the hydrogen of acetanilide ring present in para position selectively and forms *p*-bromoacetanilide.

Acetanilide $\xrightarrow[Br_2]{\text{MW 425 W, 5 min}}$ *p*-Bromoacetanilide

Chemicals required

- Acetanilide
- Bromine
- Glacial acetic acid

Procedure:

1) Load acetanilide (1.35 g, 10 mmol) and add bromine solution drop wise to the microwave flask (beaker, 50 ml) from dropping funnel.
2) (prepare the bromine solution (0.53 ml) in glacial acetic acid (2.5 ml) and transfer it into a dropping funnel)
3) Place the flask into the Microwave oven and switch on the magnetic stirrer.
    - In case of Microwave synthesizer, attach the reflux condenser to the flask.
    - In case of domestic microwave oven, cover the flask with a petridish loaded with ice cubes (provides condensing effect), alternatively funnel can be used. Beaker containing cold water can be used as a heating sink.
4) Irradiate the reaction mixture present in reaction flask with Microwaves at 50% intensity for 5 minutes, with intermittent cooling for one minute after 3 minutes of microwave irradiations.
5) Turn off the power supply to the Microwave oven.
6) After few minutes, remove the flask from the Microwave oven.
7) Cool the reaction mixture to room temperature and pour the mixture into beaker containing cold water (50 ml).
8) Filter the yellowish precipitate under suction, wash with cold water and dry.
9) Recrystallize the pure *p*-bromoacetanilide from ethanol.

Note: Monitor the progress of reaction through thin-layer chromatography (TLC).

Comparison of conventional and microwave method

	Conventional	*Microwave*
Reaction time	45 min	5 min
Reaction temperature	30-40 °C	50% intensity (425 W)
Colour	White	
Melting point	167-168 °C	
Percentage yield	76%	80%

Post Lab Questions:

1) What is bromination?
2) What are the electrophile and nucleophile generated during this reaction.
3) Why *p*-bromoacetanilide is major product.

ALKYLATION

Alkyaltion involves the replacement of active hydrogen of molecular functional groups by an alkyl group of alkylating agent. The functional groups include hydroxyl (*O*-alkylation), amino (1° and 2°; N-alkylation) and sulphhydryl (S-alkylation) groups. Alkyation with one carbon alkylating agent is called as methylation.

EXPERIMENT 11

Preparation of Methyl-2-Naphthyl Ether

Principle: Methyl-2-naphthyl ether is also called as nerolin. Preparation of methyl 2-naphthyl ether involves O-alkylation reaction. Alkylation of 2-naphthol with dimethyl sulphate in alkaline condition (sodium hydroxide) forms methylated-2-naphthol (ether). This product is known as methyl-2-naphthyl ether.

OH MW 425 W, 10 min $\xrightarrow{}$ $(CH_3)_2SO_4$ NaOH OCH_3

2-Naphthol → Methyl-2-Naphthyl ether

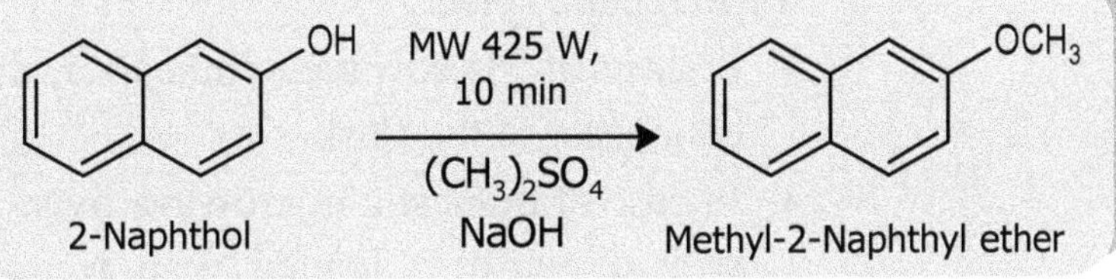

Reaction mechanism

Formation of nucleophile

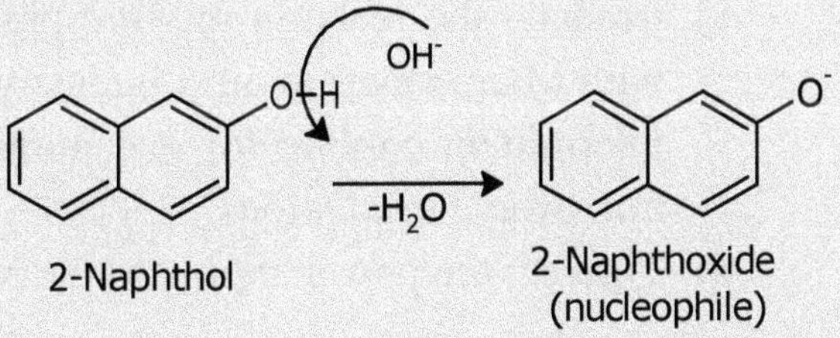

O–H OH^- $-H_2O$ O^-

2-Naphthol → 2-Naphthoxide (nucleophile)

Methylation of nucleophile (generation of methyl-2-naphthyl ether)

2-Naphthoxide (nucleophile) + Dimethyl sulphate ($H_3C-O-SO_2-O-CH_3$) → Methyl-2-Naphthyl ether (OCH_3)

Chemicals required

- 2-naphthol (β-naphthol)
- Sodium hydroxide
- Dimethyl sulphate

Procedure:

1) Load 2-naphthol (1.4 g, 9 mmol) into microwave flask (beaker, 50 ml) and dissolve it with the solution of sodium hydroxide (0.4 g in 7 ml of water).
2) Place it on the ice-bath to reduce the temperature below 10 °C (exothermic reaction).
3) Add dimethyl sulphate (0.94 ml) drop wise with vigorous stirring.
4) Place the flask into the Microwave oven and switch on the magnetic stirrer.
    - In case of Microwave synthesizer, attach the reflux condenser to the flask.
    - In case of domestic microwave oven, cover the flask with a petridish loaded with ice cubes (provides condensing effect), alternatively funnel can be used. Beaker containing cold water can be used as a heating sink.
5) Irradiate the reaction mixture present in reaction flask with Microwaves at 50% intensity for 10 minutes, with intermittent cooling for one minute after 3 minutes of microwave irradiations.
6) Turn off the power supply to the Microwave oven.

7) After few minutes, remove the flask from the Microwave oven.
8) Cool the reaction mixture and filter.
9) Wash the filter with sodium hydroxide (10%) followed by water.
10) Dry the washed residue and purify through recrystallization with industrial sprit.

Note: Monitor the progress of reaction through thin-layer chromatography (TLC).

Comparison of conventional and microwave method

	Conventional	***Microwave***
Reaction time	1 h	10 min
Reaction temperature	70-80 °C	50% intensity (425 W)
Colour	White	
Melting point	72-73 °C	
Percentage yield	65%	78%

Post Lab Questions:

1) What is called as alkylation reaction?
2) What are the common alkylating agents?
3) Mention the pharmaceutical importance of alkylation.

EXPERIMENT 12

Preparation of Anisole

Principle: Methylation of phenols with dimethyl sulphate under alkaline conditions produce methyl-phenyl ether. This reaction is typical example for *O*-alkylation.

Phenol → (MW 425 W, 5 min; $(CH_3)_2SO_4$, K_2CO_3, acetone) → Methyl-Phenylether (OCH_3)

Chemicals required

- Dimethyl sulphate
- Potassium carbonate (anhydrous)
- Acetone
- Hydrochloric acid (10%)
- Sodium carbonate solution (10%)
- Sodium chloride solution
- diethyl ether

Procedure:

1) Load phenol (1 ml, 11.4 mmol), dry dimethyl sulphate (0.15 g), anhydrous potassium carbonate (0.4 g), dry acetone into microwave flask (beaker, 50 ml) and mix well.
2) Place the flask into the Microwave oven and switch on the magnetic stirrer.
   - In case of Microwave synthesizer, attach the reflux condenser to the flask.
   - In case of domestic microwave oven, cover the flask with a petridish loaded with ice cubes (provides condensing effect), alternatively funnel can be used. Beaker containing cold water can be used as a heating sink.
3) Irradiate the reaction mixture present in reaction flask with Microwaves at 50% intensity for 5 minutes, with intermittent cooling for one minute after 3 minutes of microwave irradiations.
4) Turn off the power supply to the Microwave oven.
5) After few minutes, remove the flask from the Microwave oven.
6) Pour the contents into a beaker containing ice water.
7) Neutralize the solution with hydrochloric acid (10%) and extract the product with diethyl ether (25 ml, three times).

8) Combine the ether extracts, wash it with solution of sodium bicarbonate solution (10%) and sodium chloride solution
9) Dry the resulting solution over anhydrous sodium sulphate (filter) and evaporate the solvent ether to get colourless liquid of anisole.

Note: Monitor the progress of reaction through thin-layer chromatography (TLC).

Comparison of conventional and microwave method

	Conventional	***Microwave***
Reaction time	1 h	5 min
Reaction temperature	50-60 °C	50% intensity (425 W)
Colour	Colourless	
Melting point	154 °C	
Percentage yield	64%	68%

Post Lab Questions:

1) What is *O*-alkylation?
2) What is the role of the potassium carbonate in this reaction?
3) Why drying over anhydrous sodium sulphate is required?

BENZILIC ACID REARRANGEMENT

The base catalyzed conversion of benzil into the benzilic acid (α-hydroxy carboxylic acid) is known as benzilic acid rearrangement. The most commonly used base in this reaction is alcoholic potassium hydroxide (sodium ethoxide also useful).

EXPERIMENT 13

Preparation of Benzilic Acid

Principle: Benzil can be converted into benzilic acid in presence of alcoholic potassium hydroxide. This chemical conversion involves Benzilic acid rearrangement. Ethoxide anion attacks the one of the carbonyl carbon of benzil and gets converted into Benzilic acid.

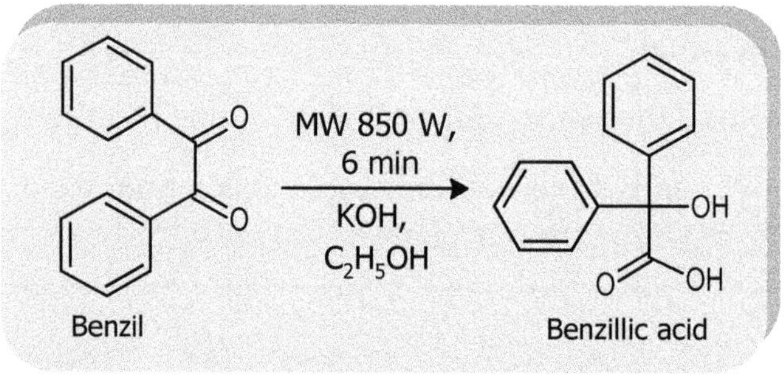

O
O
MW 850 W,
6 min
KOH,
C_2H_5OH
OH
O
OH
Benzil
Benzillic acid

Reaction mechanism

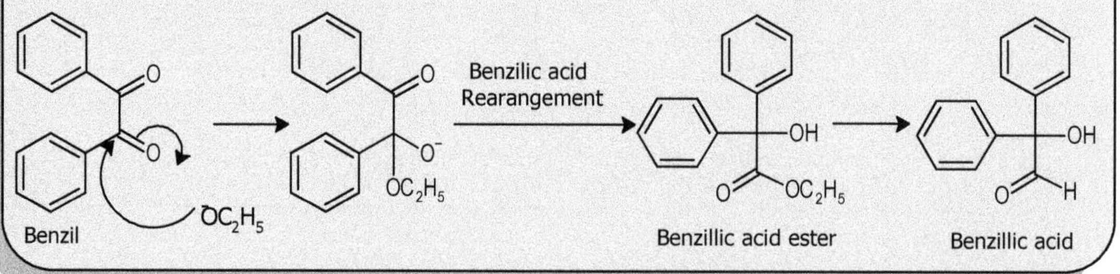

O
O
OC_2H_5
Benzil
O
O^-
OC_2H_5
Benzilic acid
Rearangement
OH
O
OC_2H_5
Benzillic acid ester
OH
O
H
Benzillic acid

Chemicals required

- Benzil
- Ethanol
- Alcoholic potassium hydroxide solution (20%)
- Dilute sulphuric acid.s

Procedure:

1) Load benzil (0.5 g, 2.4 mmol) and alcoholic potassium hydroxide (6 ml; 20%) into the microwave flask (beaker, 50 ml) and mix it well to dissolve the contents.
2) Place the flask into the Microwave oven and switch on the magnetic stirrer.
   - In case of Microwave synthesizer, attach the reflux condenser to the flask
   - In case of domestic microwave oven, cover the flask with a petridish loaded with ice cubes (provides condensing effect), alternatively funnel can be used. Beaker containing cold water can be used as a heating sink.
3) Irradiate the reaction mixture present in reaction flask with Microwaves at 100% intensity for 6 minutes, with intermittent cooling for one minute after 3 minutes of microwave irradiations.
4) Turn off the power supply to the Microwave oven.
5) After few minutes, remove the flask from the Microwave oven.
6) Cool the conical flask to get the fine crystals of potassium benzilate.
7) Re-dissolve the product in distilled water (20 ml) and filter to remove any insoluble impurities.
8) Add dilute sulphuric acid (20 ml) to the clear filtrate until separation of acid is complete.
9) Cool, filter and wash the precipitate with hot distilled water and dry.
10) Recrystallize if necessary with ethanol.

Note: Monitor the progress of reaction through thin-layer chromatography (TLC).

Comparison of conventional and microwave method

	Conventional	***Microwave***
Reaction time	30 min	6 min
Reaction temperature	50-60 °C	100% intensity (850 W)
Colour	White	
Melting point	151-152 °C	
Percentage yield	85%	90%

Post Lab Questions:

1) What is rearrangement reaction?
2) What is the role of potassium hydroxide and sulphuric acid in this reaction?
3) Explain the reaction mechanism involved in the Benzilic acid rearrangement.

CLAISEN-SCHMIDT CONDENSATION

Aldehyde and ketone condenses and generates β-hydroxyl carbonyl compound by eliminating water molecule. The condensation of a ketone containing α-hydrogen with an aldehyde that has no α-hydrogen under aqueous / alcoholic acid or base is known as Claisen-Schmidt condensation. This kind of condensation of different carbonyl compounds are known as crossed-aldol-condensation reaction. It generate, α, β-unsaturated carbonyl system. The condensation of benzaldehyde and acetophenone is an example. This condensation reaction generates benzylidene acetophenone.

Benzaldehyde + Acetophenone $\xrightarrow{H^+/OH^-}$ Benzylidene acetophenone

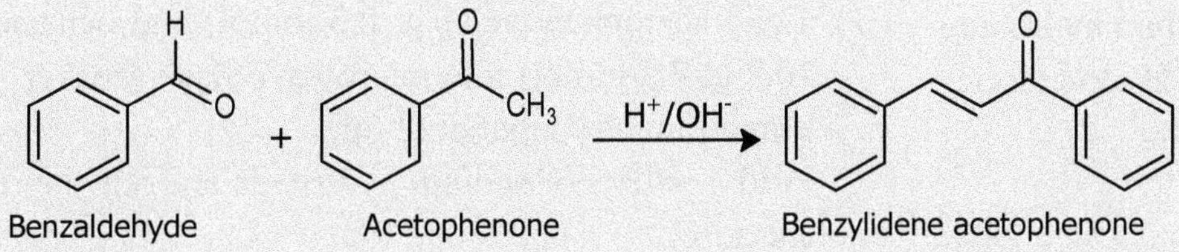

EXPERIMENT 14

Preparation of Benzylidene Acetophenone

Principle: Benzaldehyde condenses with acetophenone and generates benzylidene acetophenone. This molecule belongs to the chemical category of chalcones. This reaction is mostly catalyzed by the ethanolic sodium hydroxide.

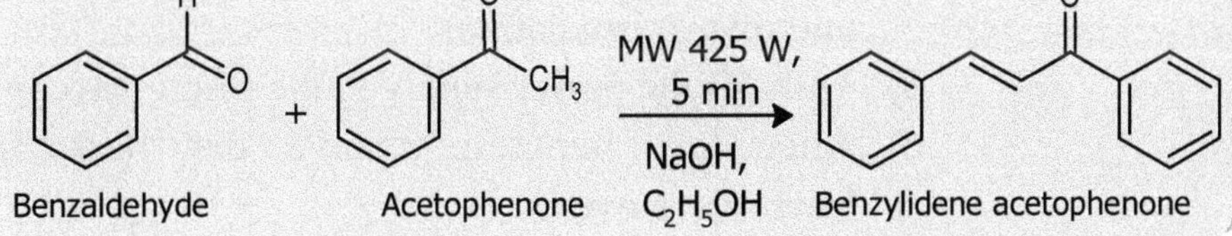

Benzaldehyde + Acetophenone $\xrightarrow[NaOH,\ C_2H_5OH]{MW\ 425\ W,\ 5\ min}$ Benzylidene acetophenone

Reaction mechanism:

Benzaldehyde | H_2C^- | Acetophenone | ⇌ | OH O | H | β-Hydroxy carbonyl compound | → $-H_2O$ | Benzylidene acetophenone (chalcone)

Chemicals required

- Benzaldehyde
- Acetophenone
- Sodium hydroxide and Hydrochloric acid

Procedure:

1) Load acetophenone (1 g, 8.3 mmol) and benzaldehyde (0.8 g, 7.5 mmol) into microwave flask (beaker, 50 ml) containing dry ethanol (2 ml).
2) Add 2 pellets of sodium hydroxide and stir the reaction mixture.
3) Place the flask into the Microwave oven and switch on the magnetic stirrer.
    - In case of Microwave synthesizer, attach the reflux condenser to the flask.
    - In case of domestic microwave oven, cover the flask with a petridish loaded with ice cubes (provides condensing effect), alternatively funnel can be used. Beaker containing cold water can be used as a heating sink.
4) Irradiate the reaction mixture present in reaction flask with Microwaves at 50% intensity for 5 minutes, with intermittent cooling for one minute after 3 minutes of microwave irradiations.
5) Turn off the power supply to the Microwave oven.
6) After few minutes, remove the flask from the Microwave oven.

7) Cool the reaction mixture and filter.
8) Wash it with cold ethanol followed by water till the washings are neutral and dry.
9) Recrystallize the benzylidene acetophenone from rectified spirit.

Note: Monitor the progress of reaction through thin-layer chromatography (TLC).

Comparison of conventional and microwave method

	Conventional	*Microwave*
Reaction time	3 hours	5 min
Reaction temperature	50-60 °C	50% intensity (425 W)
Colour	White	
Melting point	56-57 °C	
Percentage yield	70%	86%

Post Lab Questions:

1) What is the role of sodium hydroxide in this condensation?
2) What is chalcone?

EXPERIMENT 15

Preparation of Benzalacetone

Principle: Benzaldehyde condenses with acetone and generates benzalacetone. This molecule belongs to the chemical category of chalcones. This reaction is catalyzed by the sodium hydroxide. The reaction involves the nucleophilic addition followed by removal of water molecule. This reaction is known as Claisen-Schmidt reaction.

Benzaldehyde + Acetone $\xrightarrow[\text{NaOH, } C_2H_5OH]{\text{MW 425 W, 2 min}}$ Benzalacetone

Chemicals required

- Benzaldehyde
- Acetone
- Sodium hydroxide
- Water and Ethanol

Procedure:

1) Load acetone (0.6 ml, 8.2 mmol) and benzaldehyde (1.1 ml, 11 mmol) microwave flask (beaker, 50 ml) containing ethanol (8 ml) and water (10 ml).
2) Add 2 pellets of sodium hydroxide
3) Place the flask into the Microwave oven and switch on the magnetic stirrer.
    - In case of Microwave synthesizer, attach the reflux condenser to the flask.
    - In case of domestic microwave oven, cover the flask with a petridish loaded with ice cubes (provides condensing effect), alternatively funnel can be used. Beaker containing cold water can be used as a heating sink.
4) Irradiate the reaction mixture present in reaction flask with Microwaves at 50% intensity for 2 minutes.
5) Turn off the power supply to the Microwave oven.
6) After few minutes, remove the flask from the Microwave oven.
7) Cool the reaction mixture, filter, wash with water and dry.
8) Recrystallize the benzalacetone from rectified spirit.

Note: Monitor the progress of reaction through thin-layer chromatography (TLC).

Comparison of conventional and microwave method

	Conventional	***Microwave***
Reaction time	1 hour	2 min
Reaction temperature	25-35 °C	50% intensity (425 W)
Colour	Pale yellow	
Melting point	40-42 °C	
Percentage yield	94%	96%

Post Lab Questions:

1) What is the role of sodium hydroxide in this condensation?
2) What is chalcone?
3) Explain nucleophilic addition reactions of aldehydes and ketones.

WILLIAMSON ETHER SYNTHESIS

In the Williamson ether synthesis, an alkoxide reacts with an alkyl halide to give an ether.

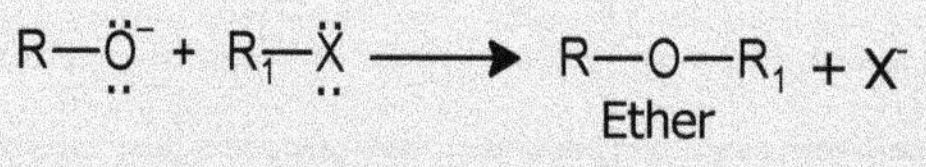

$$R-\ddot{\underset{..}{O}}^{-} + R_1-\ddot{\underset{..}{X}} \longrightarrow \underset{\text{Ether}}{R-O-R_1} + X^{-}$$

This reaction is an important example for S_N2 reaction. In this reaction the conjugate base of alcohol (or thiol) acts as a nucleophile. This nucleophile displaces the halide group of alkyl halide and forms ether.

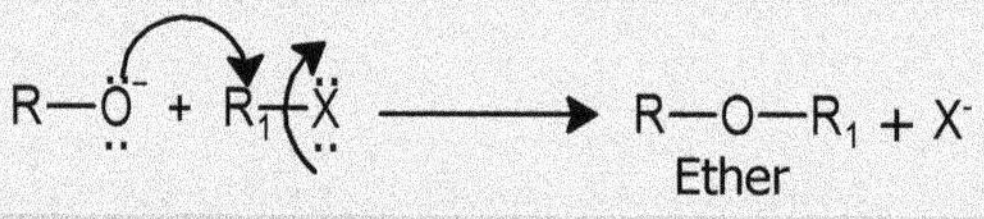

$$R-\ddot{\underset{..}{O}}^{-} + R_1-\ddot{\underset{..}{X}} \longrightarrow \underset{\text{Ether}}{R-O-R_1} + X^{-}$$

EXPERIMENT 16

Preparation of Phenacetin

Principle: Phenacetin can be prepared from paracetamol through Williamson ether reaction. Alkylation (O-alkylation) of phenol with ethyl iodide under basic condition (sodium ethoxide) forms phenacetin.

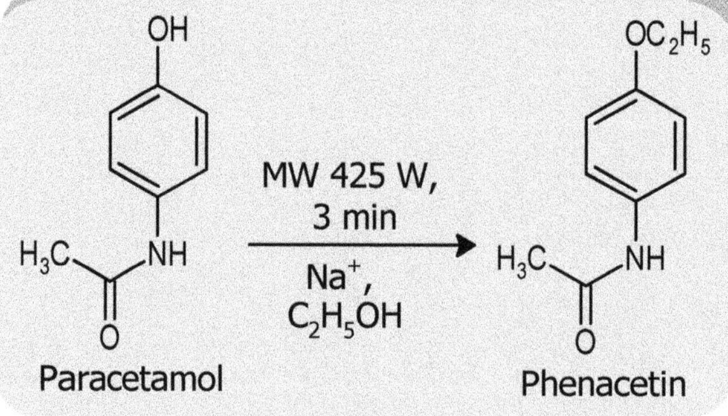

OH
H$_3$C NH O
Paracetamol
MW 425 W, 3 min
Na$^+$, C_2H_5OH
OC$_2$H$_5$
H$_3$C NH O
Phenacetin

Reaction mechanism

Chemicals required

- Paracetamol
- Ethyl Iodide
- Sodium metal
- Absolute ethanol

Procedure:

1) Dissolve sodium metal (0.1 g) in absolute ethanol (4 ml) into microwave flask (beaker, 50 ml) and stir the mixture for 10 minutes.
2) Load paracetamol (0.5 g, 3.3 mmol) and ethyl iodide (1 ml, 12.4 mmol) into the flask and mix well.
3) Place the flask into the Microwave oven, and switch on the magnetic stirrer.
   - In case of Microwave synthesizer, attach the reflux condenser to the flask.
   - In case of domestic microwave oven, cover the flask with a petridish loaded with ice cubes (provides condensing effect), alternatively funnel can be used. Beaker containing cold water can be used as a heating sink.
4) Irradiate the reaction mixture present in reaction flask with Microwaves at 50% intensity for 3 minutes.
5) Turn off the power supply to the Microwave oven.

6) After few minutes, remove the flask from the Microwave oven.
7) Cool, filter and wash the residue with cold water and dry well.
8) Purify the phenacetin by recrystallization with ethanol.

Note: Monitor the progress of reaction through thin-layer chromatography (TLC).

Comparison of conventional and microwave method

	Conventional	*Microwave*
Reaction time	1 hour	3 min
Reaction temperature	60-70 °C	50% intensity (425 W)
Colour	White	
Melting point	133-134 °C	
Percentage yield	84%	90%

Post Lab Questions:

1) Why ethylation (alkylation) is preferred at hydroxyl group over secondary amino group?
2) Explain the reaction mechanism for Williamson ether synthesis.

FISCHER INDOLE SYNTHESIS

Fischer indole synthesis is the important organic process to produce substituted indoles. Fischer indole synthesis involves the following two steps:

Step 1: Condensation of carbonyl compounds (aldehyde/ketone) with arylhydrazine to form arylhydrazones.

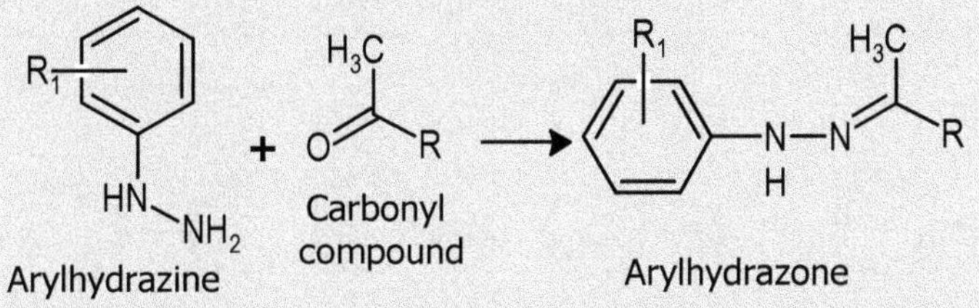

Arylhydrazine + Carbonyl compound → Arylhydrazone

Step 2; An acid catalyzed cyclization of arylhydrazone produce substituted indole.

Arylhydrazone $\xrightarrow{H^+}$ 2-Substituted indole

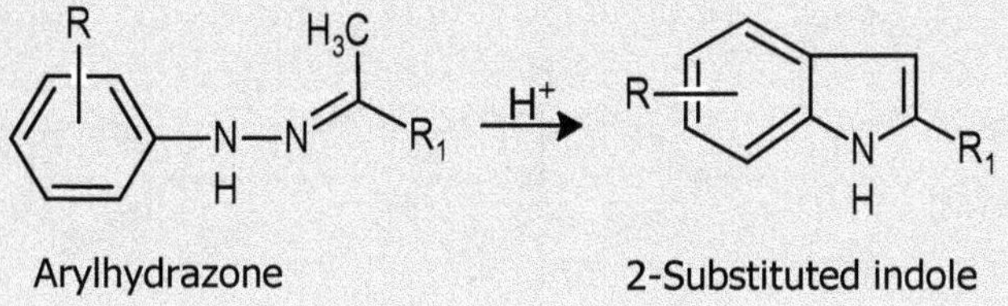

EXPERIMENT 17

Preparation of 2-Phenylindole

Principle: Acetpophenone condenses with phenylhydrazine through glacial acetic acid catalyzed reaction. This reaction is known as Fischer's indole synthesis and produces 2-phenylindole. The crude 2-phenylindole can be purified by recyrstallization with ethanol.

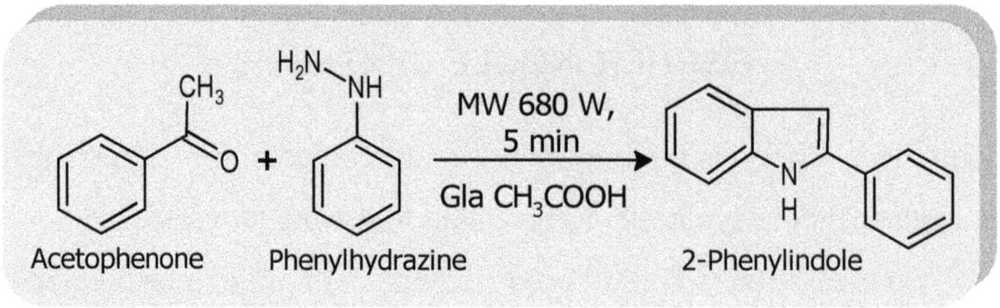

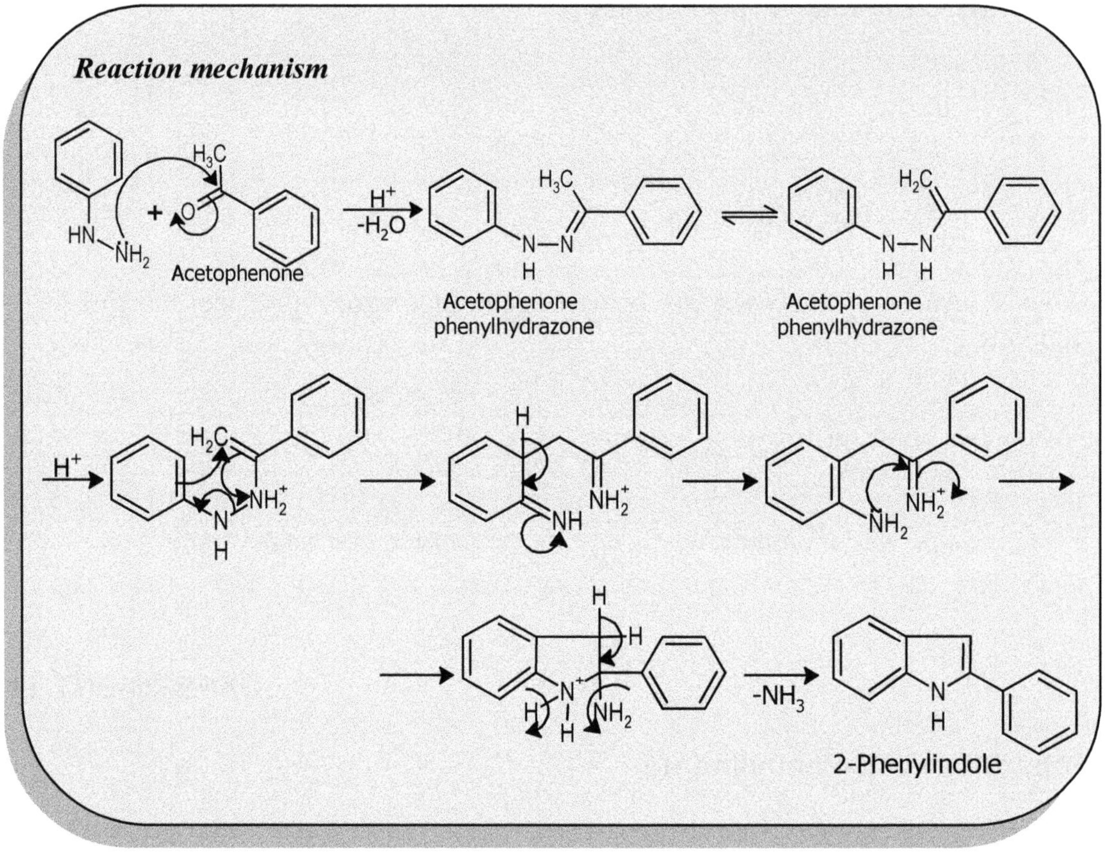

Acetophenone + Phenylhydrazine $\xrightarrow[\text{Gla } CH_3COOH]{\text{MW 680 W, 5 min}}$ 2-Phenylindole

Reaction mechanism

Acetophenone; H^+, $-H_2O$; Acetophenone phenylhydrazone ⇌ Acetophenone phenylhydrazone; H^+; $-NH_3$; 2-Phenylindole

Chemicals required

- Phenylhydrazine
- Acetophenone
- Glacial Acetic acid
- Diethyl ether
- Sodium hydroxide (2N)

Procedure:

1) Place a mixture of phenylhydrazine (1 g, 9.2 mmol), acetophenone (1.1 g, 9.2 mmol) and glacial acetic acid (10 ml) into microwave flask (beaker, 50 ml).

- Diethyl ether
- Sodium sulphate (anhydrous)

2) Place the flask into the Microwave oven and switch on the magnetic stirrer.
   - Cover the flask with a petridish loaded with ice cubes (provides condensing effect), alternatively funnel can be used. Beaker containing cold water can be used as a heating sink.
3) Irradiate the reaction mixture present in reaction flask with Microwaves at 80% intensity for 5 minutes with intermittent cooling for one minute after 3 minutes of microwave irradiations.
4) Turn off the power supply to the Microwave oven.
5) After few minutes, remove the flask from the Microwave oven.
6) Cool to room temperature, pour the contents slowly into the beaker containing ice-cold water (100 ml).
7) Filter the solid obtained and dissolve it in solvent diethyl ether.
8) Wash the ether layer with sodium hydroxide (2N, 2 x 15 ml), then with water (2 x 15 ml) and finally with brine (15 ml).
9) Dry the ether layer over anhydrous sodium sulphate and evaporate to obtain the crude 2-phenyl indole.
10) Purify the crude 2-phenylindole with the mixture of ether and petroleum ether (10 ml each).

Note: Monitor the progress of reaction through thin-layer chromatography (TLC).

Comparison of conventional and microwave method

	Conventional	***Microwave***
Reaction time	3 h	5 min
Reaction temperature	80 °C	80% intensity (680 W)
Colour	White	
Melting point	190-191 °C	
Percentage yield	82%	90%

Post Lab Questions:

1) What is the role of solvent ether in the 2-phenylindole preparation?
2) What are the different acid catalysts useful in the Fischer indole synthesis?

PERKIN REACTION

Aromatic aldehydes condenses with aliphatic anhydrides (containing two α hydrogen atoms) in presence of the basic catalyst (eg: sodium acetate) and generates β-arylacrylic acid (α, β-unsaturated carboxylic acid). This condensation reaction is named as Perkin reaction.

$$Ar{-}CHO + R{-}CO{-}O{-}CO{-}R \xrightarrow{CH_3COONa} Ar{-}\overset{\beta}{C}H{=}\overset{\alpha}{C}H{-}COOH$$

Arylaldehyde; Aliphatic anhydride; α,β-Unsaturated carboxylic acid (α,β-arylacrylic acid)

EXPERIMENT 18

Preparation of Cinnamic Acid

Principle: The Perkin condensation of benzaldehyde with acetic anhydride in the presence of sodium acetate gives cinnamic acid. Initially, the reaction forms an acetyl ester of carbocyclic acid, which upon hydrolysis (base) liberates cinnamic acid. Potassium acetate can be used in the place of sodium acetate.

$$C_6H_5{-}CHO + H_3C{-}CO{-}O{-}CO{-}CH_3 \xrightarrow[CH_3COONa]{MW\ 765\ W,\ 10\ min} C_6H_5{-}CH{=}CH{-}COOH + CH_3COOH + H_2O$$

Benzaldehyde; Acetic anhydride; Cinnamic acid

The sodium acetate (acyloxy ion) acts as a base and abstracts the proton from methyl group of acetic anhydride and generates nucleophile. Nucleophile of acetic anhydride attacks positively charged carbon (due to positive inductive effect) of carbonyl group in benzaldehyde and forms intermediate. This intermediate on heating eliminates water molecule and on hydrolysis removes acetate to yield cinnamic acid.

Reaction mechanism

CH_3COONa → CH_3COO^- + Acetic anhydride → Nucleophile + CH_3COOH

Benzaldehyde + Nucleophile ⇌ ($-H_2O$) → (Na_2CO_3, HCl, $-CH_3COOH$) → C_6H_5-CH=CH-COOH (Cinnamic acid)

Chemicals required

- Benzaldehyde
- Acetic anhydride
- Sodium acetate and Hydrochloric acid

Procedure:

1) Place benzaldehyde (1.05 g, 1 ml, 9.8 mmol), acetic anhydride (1.5 g, 1.4 ml, 15 mmol) and fused and finely powdered sodium acetate (0.6 g) into mortor. Grind well and transfer into microwave flask (beaker, 50 ml).
2) Place the flask into the Microwave oven and switch on the magnetic stirrer.
   - In case of Microwave synthesizer, attach the reflux condenser to the flask.
   - Cover the flask with a petridish loaded with ice cubes (provides condensing effect), alternatively funnel can be used. Beaker containing cold water can be used as a heating sink.
3) Irradiate the reaction mixture present in reaction flask with Microwaves at 90% intensity for 10 minutes with intermittent cooling for one minute after 3 minutes of microwave irradiations.

4) Turn off the power supply to the Microwave oven.
5) After few minutes, remove the flask from the Microwave oven.
6) Cool to room temperature; pour the hot mixture into the beaker containing cold water (10 ml).
7) Add saturated sodium carbonate solution with shaking until a drop turns red litmus blue.
8) Filter at pump to remove resinous materials, if any present and acidify the filtrate by with hydrochloric acid.
9) Filter the precipitate, wash with cold water and dry.
10) Purify the product by recrystallization with hot water.

Note: Monitor the progress of reaction through thin-layer chromatography (TLC).

Comparison of conventional and microwave method

	Conventional	***Microwave***
Reaction time	4 h	10 min
Reaction temperature	50-60 °C	80% intensity (765 W)
Colour	White	
Melting point	132-133 °C	
Percentage yield	88%	92%

Post Lab Questions:

1) Why sodium acetate is used in Perkin condensation?

2) What is the product of Perkin condensation?

HYDANTOIN SYNTHESIS

The imidazolidine-2,4-dione nucleus is known as hydantoin. The hydantoin scaffold is present in variety of alkaloids such as aplysinopsins (neuro-transmission inhibitor), (E)-axinohydantoin (protein kinase-C inhibitor), mukanadin B (anti-viral) and mipacamide (anti-fungal). It has wide range of therapeutic potential. Baeyer's synthesis, Urech method, Read method, Strecker synthesis and Bucherer-Bergs reaction are useful in synthesizing hydantoin containing molecules.

EXPERIMENT 19

Preparation of Hydantoin

Principle: Glycine condenses with urea in presence of alkali to form hydantoic acid. Hydantoic acid on acid catalyzed cyclization forms hydantoin. This reaction eliminates ammonia.

$$H_2N-CO-NH_2 + H_2N-CH_2-COOH \xrightarrow[NaOH]{MW\ 595\ W,\ 5\ min} H_2N-CO-NH-CH_2\cdot COOH \xrightarrow[HCl,\ -NH_3]{MW\ 595\ W,\ 4\ min} \text{Hydantoin}$$

Urea, Glycine, Hydantoic acid, Hydantoin

Reaction mechanism

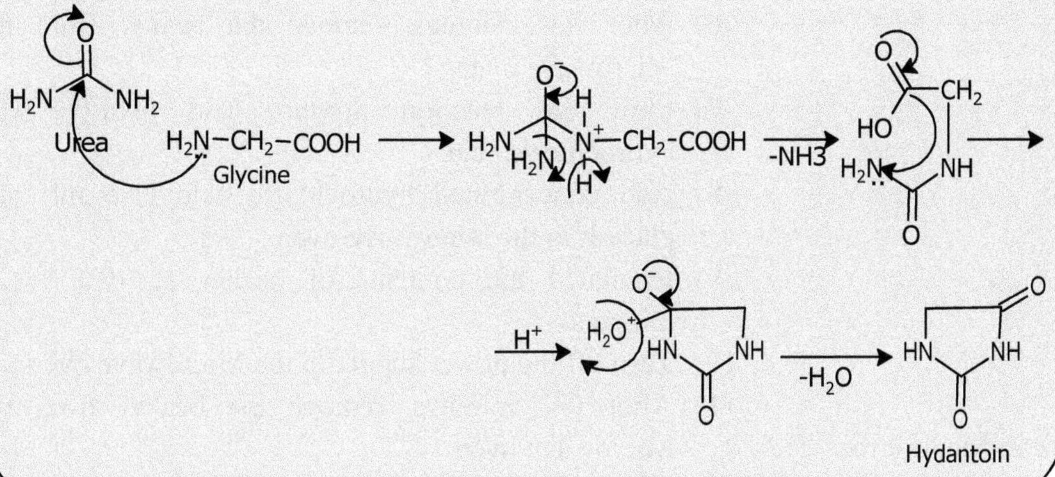

Chemicals required

- Glycine
- Urea
- Sodium hydroxide
- Concentrated hydrochloric acid

Procedure:

1) Load glycine (0.95 g, 13 mmol), urea (1.5 g, 25 mmol) and sodium hydroxide (0.5 g) into into mortor. Grind well and transfer into microwave flask (beaker, 50 ml).
2) Place the beaker into the Microwave oven and switch on the magnetic stirrer.
    - can be used. Beaker containing cold water can In case of Microwave synthesizer, attach the reflux condenser to the flask.
    - Cover the flask with a petridish loaded with ice cubes (provides condensing effect), alternatively funnel be used as a heating sink.
3) Irradiate the reaction mixture present in the beaker with Microwaves at 70% intensity for 5 minutes with intermittent cooling for two minutes after 3 minutes of microwave irradiations.

4) During cooling stir the reaction mixture with spatula.
5) Turn off the power supply to the Microwave oven.
6) After few minutes, remove the beaker from the Microwave oven.
7) Cool the reaction mixture and acidify with hydrochloric acid.
8) Add concentrated hydrochloric acid (1.5 ml) and place it in the microwave oven.
9) Irradiated the contents of beaker at 70% for 4 minutes.
10) Turn off the power supply to the Microwave oven.
11) After few minutes, remove the beaker from the Microwave oven.
12) Cool the contents by placing in the ice bath.
13) Add methanol to remove the excess of acid and filter.
14) Purify the solid by recrystallization with ethanol.

Note: Monitor the progress of reaction through thin-layer chromatography (TLC).

Comparison of conventional and microwave method

	Conventional	***Microwave***
Reaction time	1.5 h	9 min
Reaction temperature	80-90 °C	70% intensity (595 W)
Colour	White	
Melting point	218 °C	
Percentage yield	89%	90%

Post Lab Questions:

1) Give the general chemical reaction for hydantoin synthesis?
2) What are the therapeutic potentials of hydantoins? Name one drug containg hydantoin nucleus.

BILTZ SYNTHESIS

German scientist Biltz (1908) reported the straight forward synthesis of imidazolidine-2,4-dione (hydantoin) later named as phenytoin. The condensation of benzil and urea in the presence of potassium hydroxide and ethanol as solvent produced phenytoin. This reaction is known as Biltz synthesis. This reaction proceeds through Benzillic rearrangement.

Benzil + $H_2N-CO-NH_2$ (Urea) $\xrightarrow{KOH}$ Phenytoin

EXPERIMENT 20

Preparation of Phenytoin

Principle: Phenytoin can be prepared by the reaction between benzil and urea in alkaline conditions using methanol as a solvent. Reaction involves removal of water molecule and rearrangement to give rise to phenytoin, which is widely used as anticonvulscent.

Benzil + $H_2N-CO-NH_2$ (Urea) $\xrightarrow[\text{KOH, } CH_3OH\text{, HCl}]{\text{MW 595 W, 8 min}}$ Phenytoin

Reaction mechanism

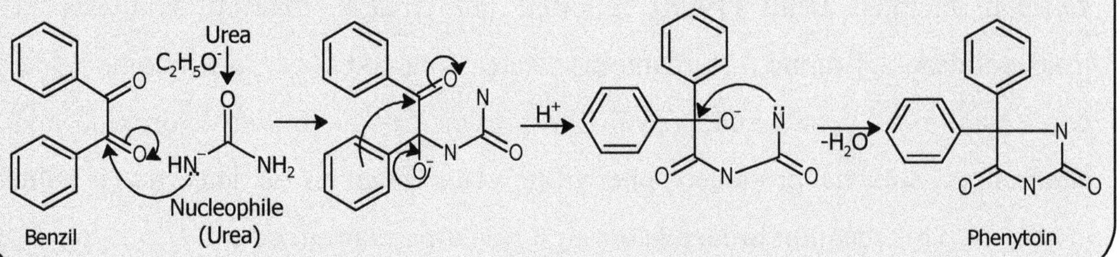

Benzil

Nucleophile (Urea)

Phenytoin

Chemicals required

- Benzil
- Urea
- Methanol
- Sodium hydroxide (30%)
- Concentrated hydrochloric acid

Procedure:

1) Load benzil (0.53 g, 2.52 mmol), urea (0.3 g, 5 mmol), aqueous sodium hydroxide (3 ml, 30%); and methanol (3 ml) into microwave flask (beaker, 50 ml).
2) Place the flask into the Microwave oven and switch on the magnetic stirrer.
    - In case of Microwave synthesizer, attach the reflux condenser to the flask.
    - Cover the flask with a petridish loaded with ice cubes (provides condensing effect), alternatively funnel can be used. Beaker containing cold water can be used as a heating sink.
3) Irradiate the reaction mixture present in the flask with Microwaves at 70% intensity for 8 minutes, with intermittent cooling for one minute after 3 minutes of microwave irradiations.
4) Turn off the power supply to the Microwave oven.
5) After few minutes, remove the flask from the Microwave oven.
6) Cool the reaction flask and dissolve the resulting solid with water (30 ml).

7) Filter and add concentrated hydrochloric acid drop wise to the filtrate.
8) Isolate the white solid and dry.
9) Purify the crude phenytoin by recrystallization with ethanol

Note: Monitor the progress of reaction through thin-layer chromatography (TLC)

Comparison of conventional and microwave method

	Conventional	*Microwave*
Reaction time	2.5 hour	8 min
Reaction temperature	78 °C	70% intensity (595 W)
Colour	White	
Melting point	297-298 °C	
Percentage yield	78%	90%

Post Lab Questions:

1) What is the role of acid in the Biltz synthesis?
2) What is Biltz synthesis?
3) What is known as aryl shift (1, 2-shift)?

SCHIFF BASES

Schiff bases are the important or ganic intermediates. The primary amines condense with a aryl or heteroayl aldehydes and generates Schiff base. Schiff base contains a carbon-nitrogen double bond (-CH=N-) and is known as azomethane and aldimines. The un-ionized amine is reactive and the amines in salt form (ionized) require deportation. The lone pair of electrons present in the amino group attacks the positively charged carbonyl carbon (polarization) and forms a carbinolamine (intermediate). The carbinolamine eliminates water molecule and forms a Schiff base. Schiff base formation is an example for nucleophilic addition followed by removal of water molecule.

$$R-CHO + H_2N-R_1 \xrightarrow[-H_2O]{\Delta} R-CH=N-R_1$$

Arylaldehyde + Primary amine → Schiff base

Reaction mechanism

$$R-CHO + H_2N-R_1 \longrightarrow R-CH(O^-)-N^+H_2-R_1 \longrightarrow R-CH(OH)-NH-R_1 \xrightarrow[-H_2O]{\Delta} R-CH=N-R_1$$

Arylaldehyde → Carbinolamine → Schiff base

EXPERIMENT 21

Preparation of Furanylmethylidene Aminophenol (Solventless Reaction)

Principle: The condensation of furfuraldehyde and 4-aminophenol generates furanylmethylidene amino phenol. This reaction can be effected suing microwave irradiation coupled with solid phase conditions (silica gel). Silica gel acts as an adsorbent and adsorb the released water molecules.

Furfuraldehyde + 4-Aminophenol → (MW 765 W, 6 min, Silica Gel) Furanylmethylidene aminophenol

Reaction mechanism

Furfuraldehyde + 4-Aminophenol → Carbinolamine

Carbinolamine → (Δ, $-H_2O$) Furanylmethylidene aminophenol

Chemicals required

- *para*-Amino phenol
- Furfuraldehyde
- Silica gel (Activated)

Procedure:

1) Load furfuraldehyde (0.96 g, 10 mmol), 4-amino phenol (1.09 g, 10 mmol) and silica gel (1 g) into a mortar.
2) Grind the mixture using pestle (trituration), transfer into microwave flask (beaker, 50 ml) and add 4 drops of dichloromethane (DCM).
3) Place the flask into the Microwave oven and switch on the magnetic stirrer.

4) Irradiate the reaction mixture present in the beaker with Microwaves at 90% intensity for 6 minutes, with intermittent cooling for 1 minute after 2 minutes microwave irradiation to avoid charring. Stir the reaction mixture with spatula, while cooling.
5) During cooling stir the reaction mixture with spatula.
6) Turn off the power supply to the Microwave oven.
7) After few minutes, remove the beaker from the Microwave oven.
8) Cool the hot mixture and extract with hot methanol and evaporate solvent to get Schiff base.
9) Purify the crude product by recrystallization with ethanol.

Note: Monitor the progress of reaction through thin-layer chromatography (TLC).

Comparison of conventional and microwave method

	Conventional	***Microwave***
Reaction time	4 h	6 min
Reaction temperature	60 °C	90% intensity (765 W)
Colour	White	
Melting point	112-113 °C	
Percentage yield	80%	82%

Post Lab Questions:

1) What is the role of silica gel in this reaction?
2) What are the alternatives to silica gel?
3) List the pharmaceutically useful Schiff bases.

EXPERIMENT 22

Preparation of Furanylmethylidene Aminophenol (Solvent Reaction)

Principle: The condensation of furfuraldehyde and 4-aminophenol generates furanylmethylidene amino phenol. This reaction can be effected using microwave irradiation coupled with solid phase conditions (silica gel). Silica gel acts as an adsorbent and adsorb the released water molecules.

Furfuraldehyde (CHO) + 4-Aminophenol (H_2N–C_6H_4–OH) —(MW 765 W, 8 min, Ethanol)→ Furanylmethylidene aminophenol (CH=N–C_6H_4–OH)

Chemicals required

- *para*-Amino phenol
- Furfuraldehyde
- Silica gel (Activated)

Procedure:

1) Load furfuraldehyde (0.96 g, 10 mmol), 4-amino phenol (1.09 g, 10 mmol) and ethanol (2 ml) into a mortar.
2) Grind the mixture using pestle (trituration), transfer into microwave flask (beaker, 50 ml) and add 4 drops of dichloromethane (DCM).
3) Place the flask into the Microwave oven and switch on the magnetic stirrer.
4) Irradiate the reaction mixture present in the beaker with Microwaves at 90% intensity for 8 minutes, with intermittent cooling for 1 minute after 2 minutes microwave irradiation to avoid charring. Stir the reaction mixture with spatula, while cooling.

5) During cooling stir the reaction mixture with spatula.
6) Turn off the power supply to the Microwave oven.
7) After few minutes, remove the beaker from the Microwave oven.
8) Cool the hot mixture and extract with hot methanol and evaporate solvent to get Schiff base.
9) Purify the crude product by recrystallization with ethanol.

Note: Monitor the progress of reaction through thin-layer chromatography (TLC).

Comparison of conventional and microwave method

	Conventional	*Microwave*
Reaction time	4 h	8 min
Reaction temperature	60 °C	90% intensity (765 W)
Colour	White	
Melting point	112-113 °C	
Percentage yield	80%	82%

Post Lab Questions:

1) What is the role of silica gel in this reaction?
2) What are the alternatives to silica gel?
3) List the pharmaceutically useful Schiff bases.

BERNTHSEN PHENOTHIAZINE SYNTHESIS

Phenothiazine is the central core in antihistamine, antipsychotic, sedative and antiemetic drugs. The reaction between diphenylamine and elemental sulphur at 250 °C forms phenothiazine via ring closure. This reaction is known as Bernthsen Phenothiazine synthesis.

Diphenylamine + S (Elemental Sulphur) $\xrightarrow{250\ °C}$ Phenothiazine

EXPERIMENT 23

Preparation of Phenothiazine

Principle: Phenothiazine can be prepared by melting diphenylamine with elemental sulphur. The reaction can be catalyzed with the use of metal halides (eg: $AlCl_3$ and $ZnCl_2$) and resublimed iodine. The elemental iodine gives better yields.

Diphenylamine + S (Elemental Sulphur) $\xrightarrow[I_2]{\text{MW 850 W, 6 min}}$ Phenothiazine

Chemicals required

- Diphenylamine
- Elemental sulphur
- Iodine

Procedure:

1) Place diphenylamine (3 g, 18 mmol) and elemental sulphur (3 g) into a mortar and triturate and transfer into microwave flask (beaker, 50 ml).
2) Add resublimed iodine (0.05 g) to the flask as a catalyst.

3) Place the flask into the Microwave oven and switch on the magnetic stirrer.
    - In case of Microwave synthesizer, attach the reflux condenser to the flask.
    - Cover the flask with a petridish loaded with ice cubes (provides condensing effect), alternatively funnel can be used. Beaker containing cold water can be used as a heating sink.
4) Irradiate the reaction mixture present in the flask with Microwaves at 100% intensity for 6 minutes with, intermittent cooling for 1 minute after 3 minutes of microwave irradiation to avoid charring.
5) During cooling stir the reaction mixture with spatula.
6) Turn off the power supply to the Microwave oven.
7) After few minutes, remove the flask from the Microwave oven.
8) Cool the molten solid and powder it using mortor and pestle.
9) Purify the product by recrystalization with ethanol.

Note: Monitor the progress of reaction through thin-layer chromatography (TLC).

Comparison of conventional and microwave method

	Conventional	***Microwave***
Reaction time	30 minutes	6 min
Reaction temperature	140-150 °C	100% intensity (850 W)
Colour	White	
Melting point	184-185 °C	
Percentage yield	68%	73%

Post Lab Questions:

1) List the drugs containing phenothiazine core structure and mention their therapeutic use.
2) What is Bernthsen Phenothiazine synthesis?

BERNTHSEN ACRIDINE SYNTHESIS

Acridines are known for antibacterial, antifungal, antimalarial and DNA intercalating potential. The reaction between diphenyl amine and carboxylic acid in presence of anhydrous zinc chloride at 200-210 °C forms phenyl acridine through ring closure. This reaction is known as Bernthsen Acridine synthesis.

COOH

N
H

+

$ZnCl_2$
$-H_2O$

Diphenylamine Benzoic acid 9-Phenylacridine

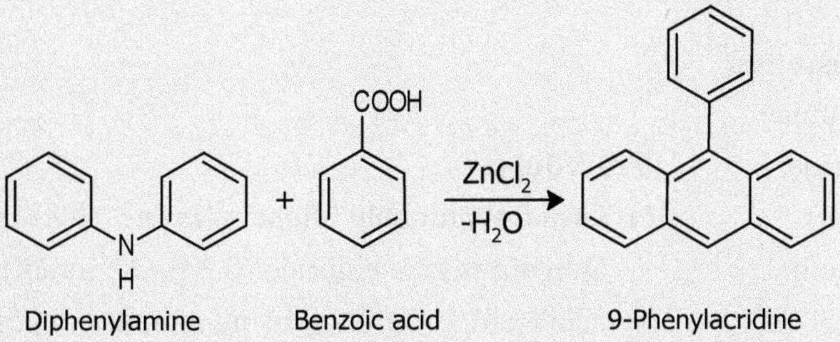

EXPERIMENT 24

Preparation of 9- Phenylacridine

Principle: 9-Phenylacridine can be prepared by heating diphenylamine with benzoic acid in presence of zinc chloride. This reaction can be effected using microwave irradiation under solid phase conditions (solvent free condition). The microwave method requires very less amount of zinc chloride compared to conventional method and produces higher yields. Hence, it serves as a typical example for green chemistry reaction.

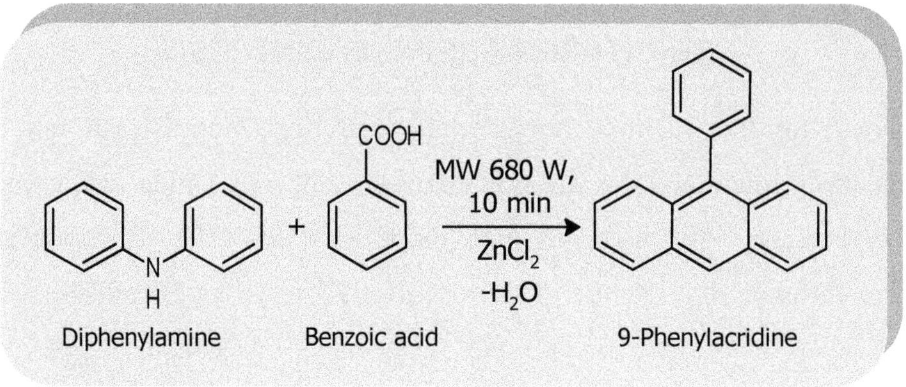

COOH

MW 680 W, 10 min

$ZnCl_2$

$-H_2O$

Diphenylamine + Benzoic acid → 9-Phenylacridine

Chemicals required

- Diphenylamine
- Zinc chloride (anhydrous)
- Benzoic acid

Procedure:

1) Grind thoroughly diphenylamine (0.83 g, 5 mmol), anhydrous zinc chloride (0.3 g, 2.2 mmol) and benzoic acid (0.61 g, 5 mmol) using mortar and pestle.
2) Transfer this reaction mixture into microwave flask (beaker, 50 ml).
3) Place the flask into the Microwave oven and switch on the magnetic stirrer.
    - In case of Microwave synthesizer, attach the reflux condenser to the flask.
    - Cover the flask with a petridish loaded with ice cubes (provides condensing effect), alternatively funnel can be used. Beaker containing cold water can be used as a heating sink.
4) Irradiate the reaction mixture present in the beaker with Microwaves at 80% intensity for 10 minutes, intermittent cooling for 1 minute, after 3 minutes of microwave irradiation to avoid charring.
5) During cooling stir the reaction mixture with spatula.
6) Turn off the power supply to the Microwave oven.

7) After few minutes, remove the beaker from the Microwave oven.
8) Cool the molten solid and powder it using mortor and pestle.
9) Purify the product by recrystalization with ethanol.

Note: Monitor the progress of reaction through thin-layer chromatography (TLC).

Comparison of conventional and microwave method

	Conventional	***Microwave***
Reaction time	3 h	10 min
Reaction temperature	100 °C	80% intensity (680 W)
Colour	White	
Melting point	184-185 °C	
Percentage yield	64%	72%

Post Lab Questions:

1) What is green chemistry?
2) What is Bernthsen Acridine synthesis?
3) Mention the role of zinc chloride in this reaction.

KNOEVENAGEL CONDENSATION

The carbonyl compounds (aldehydes and ketones) condenses with active methylene compounds (eg: ethyl acetoacetate) in presence of base catalyst and forms α, β–unsaturated compounds (-C=C- bond formation). This reaction is known as Knoevenagel condensation. Amines, sodium ethoxide, N-methyl morpholine and piperidine are most useful base catalysts. Knoevenagel condensation has significant role in fine chemical industry. Carbonyl compounds with active methylene groups present in between two electron withdrawing groups and no α-hydrogen alone undergo this reaction. The reaction proceeds through the following steps.

Step 1: Nucleophilic addition of active methylene hydrogen to the carbonyl compound.

Step 2: Dehydration of water molecules (condensation).

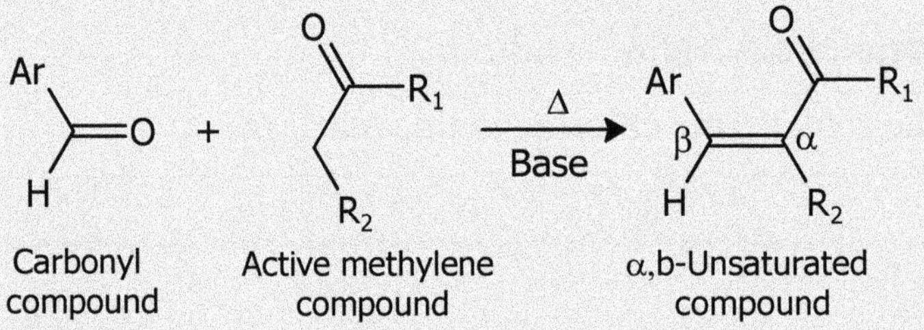

Ar, H, O, +, O, R_1, R_2, Δ, Base, Ar, β, α, H, O, R_1, R_2

Carbonyl compound | Active methylene compound | α,b-Unsaturated compound

Reaction mechanism

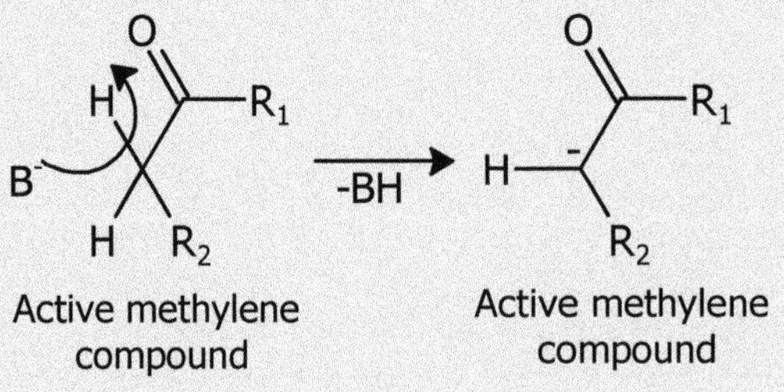

O, H, R_1, B^-, H, R_2, -BH, H, O, R_1, R_2

Active methylene compound | Active methylene compound

Ar H Carbonyl compound — Active methylene compound (R_1, R_2) → Ar–CH(O⁻)–CH(R_2)–COR_1 →(BH, -B) → OH intermediate →(-B, -BH) →(BH) → →(-B, $-H_2O$) → α,b-Unsaturated compound

EXPERIMENT 25

Preparation of Coumarin

Principle: Coumarins are additives in food and cosmetics, optical brightening agents, and dispersed fluorescent and laser dyes. Umbelliferone is very important coumarin. Anticoagulant drugs also contain coumarin nucleus (eg; warfarin). Coumarins can be synthesized by Knoevenagel condensation. Condensation of salicylaldehyde with ethyl aceto acetate in the presence of base generates coumarin. Piperidine is most preferable base for this reaction.

Salicyladehyde + Diethyl melonate →(MW 340 W, 5 min, Piperidine) 2H-1-Benzopyran-2-one (coumarin)

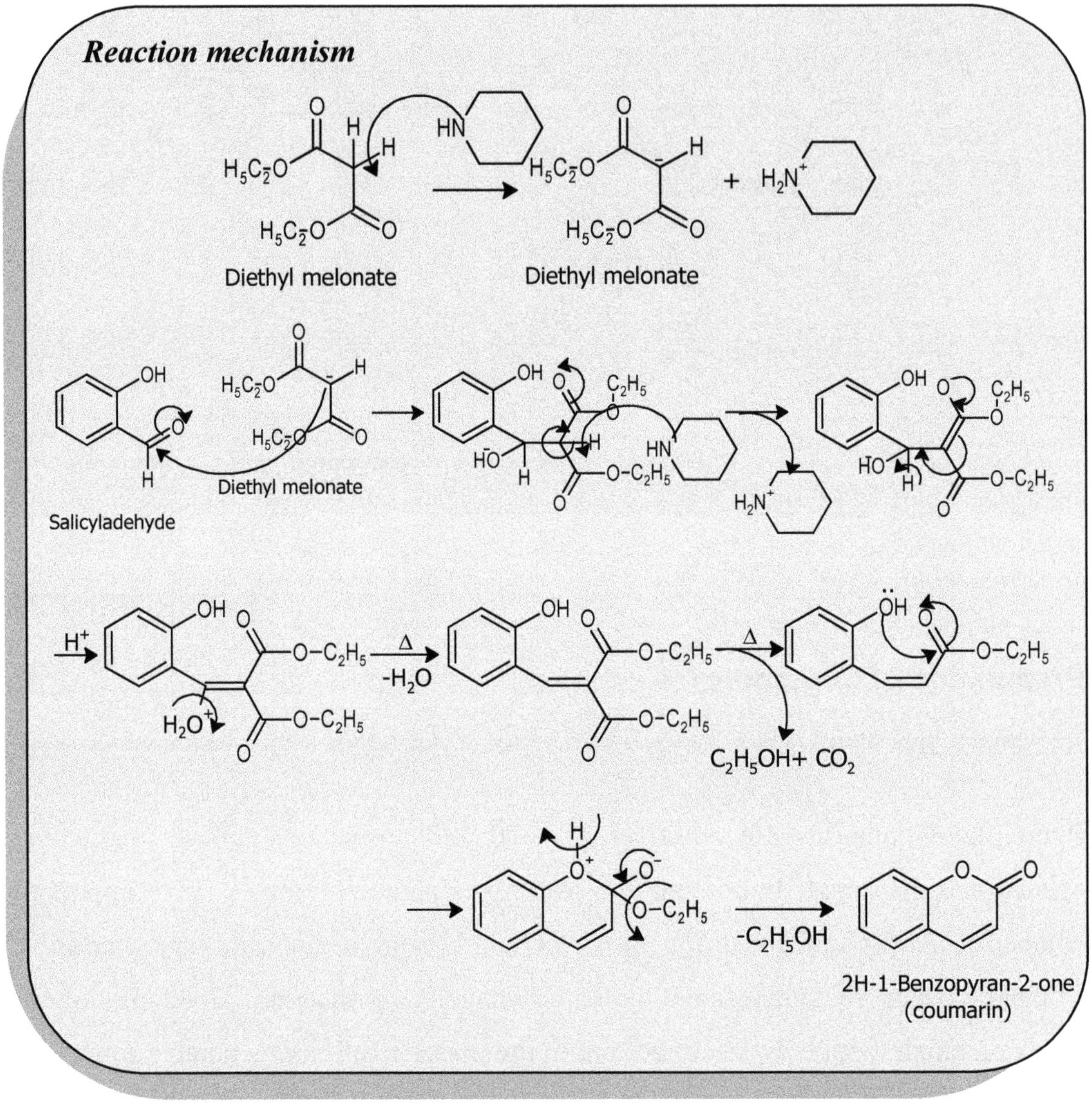

Reaction mechanism

Diethyl melonate

Diethyl melonate

Salicyladehyde

Diethyl melonate

$C_2H_5OH + CO_2$

$-C_2H_5OH$

2H-1-Benzopyran-2-one (coumarin)

Chemicals required

- Salicylaldehyde
- Diethyl malonate
- Piperidine

Procedure:

1) Load salicylaldehyde (1.22 g, 10 mmol), diethyl malonate (1.6 g, 10 mmol), and piperidine (0.1 g, 1.17 mmol) into a mortar and triturate.
2) Transfer into microwave flask (beaker, 50 ml).

3) Place the flask/beaker into the Microwave oven and switch on the magnetic stirrer.
   - In case of Microwave synthesizer, attach the reflux condenser to the flask.
   - Cover the flask with a petridish loaded with ice cubes (provides condensing effect), alternatively funnel can be used. Beaker containing cold water can be used as a heating sink.
4) Irradiate the reaction mixture present in the beaker with Microwaves at 40% intensity for 5 minutes.
5) Turn off the power supply to the Microwave oven.
6) After few minutes, remove the beaker from the Microwave oven.
7) Cool the reaction mixture to room temperature.
8) Purify the product by recrystallization with ethanol.

Note: Monitor the progress of reaction through thin-layer chromatography (TLC).

Comparison of conventional and microwave method

	Conventional	***Microwave***
Reaction time	3 h	5 min
Reaction temperature	25-30 °C	40% intensity (340 W)
Colour	Colurless crystals	
Melting point	68-71 °C	
Percentage yield	90%	90%

Post Lab Questions:

1) What is the role of piperidine in this reaction?
2) Predict the alternatives for piperidine.

EXPERIMENT 26

Preparation of 5-Arylidenethiazolidine-2,4-dione

Principle: Glitazones with thiazolidine-2,4-dione nucleus are the novel class of antidiabetic agents (e.g. rosiglitzone). Thiazolidine-2,4-dione can be prepared by the condensation of chloroacetic acid and thiourea in aqueous medium under microwave conditions. The conventional method of thiazolidine-2,4-dione requires hydrochloric acid as a catalyst. Arylidene thiazolidinedione and its derivatives are an important synthon in the pharmaceutical discovery. The condensation of thiazolidine-2,4-dione and aryl/heteroaryl aldehydes in the presence of base (preferable piperidine) generates 5- arylidene thiazolidine diones. This reaction is example for Knoevenagel condensation reaction. The conventional method requires Dean-Stark apparatus to provide azeotrophic condition (remove water). But the Microwave method proceeds without the use of this apparatus. In the experiment mentioned below, benzaldehyde is condensed with thiazolidine-2,4-dione in presence of piperidine base and produced 5-benzylidene thiazolidine-2,4-dione. This reaction is performed in solid phase (siliga gel) and produced better yield.

Step 1: Preparation of thiazolidine-2,4-dione

Thiourea + Cl—CH_2—COOH (Chloroacetic acid) $\xrightarrow[H_2O]{\text{MW 255 W, 6 min}}$ Thiazolidine-2,4-dione

Step 2: Preparation of 5-benzylidene-1,3-thiazolidine-2,4-dione

Thiazolidine-2,4-dione + Benzaldehyde $\xrightarrow[\text{15 min}]{\text{MW 850 W}}$ 5-Benzylidene-1,3-thiazolidine-2,4-dione + H_2O

Reaction mechanism

Piperidine + Thiazolidine-2,4-dione → H_2N^+ (piperidinium) + carbanion (HC^-)

Benzaldehyde + Thiazolidine-2,4-dione (carbanion) → alkoxide intermediate (O^-) $\xrightarrow[-H_2O]{\Delta}$ 5-Benzylidene-1,3-thiazolidine-2,4-dione

Step 1: ***Preparation of thiazolidine-2,4-dione***

Chemicals required

- Thiourea
- Chloroacetic acid
- Water

Procedure:

1) Load chloroacetic acid (0.95 g, 10 mmol), thiourea (0.76 g, 10 mmol) and water (15 ml) into microwave flask (beaker, 50 ml) and mix it well (~6 min; 5-10 °C).
2) Place the flask/beaker into the Microwave oven and switch on the magnetic stirrer.
   - In case of Microwave synthesizer, attach the reflux condenser to the flask.
   - Cover the flask with a petridish loaded with ice cubes (provides condensing effect), alternatively funnel can be used. Beaker containing cold water can be used as a heating sink.
3) Irradiate the reaction mixture present in the beaker with Microwaves at 30% intensity for 6 minutes, with intermittent cooling (to avoid charring) for one minute after every two minutes of microwave irradiation.
4) Turn off the power supply to the Microwave oven.
5) After few minutes, remove the beaker from the Microwave oven.
6) Cool and add cold water (5 ml) to the reaction mixture.
7) Filter the solid and purify the product by recrystalization with water.
8) The melting range of the product is 125-127 °C.

Comparison of conventional and microwave method

	Conventional	*Microwave*
Reaction time	12 h	6 min
Reaction temperature	100 °C	30% intensity (255 W)
Colour	White crystals	
Melting point	125-127 °C	
Percentage yield	88%	90%

Step 2: ***Preparation of 5-benzylidene-1,3-thiazolidine-2,4-dione***
METHOD 1 ***(solvent less reaction)***

Chemicals required

- Thiazolidine-2,4-dione
- Benzaldehyde
- Piperidine
- Acetic acid
- Dichloromethane
- Activated silica gel

Procedure:

1) Load thiazolidine-2,4-dione (0.117 g, 1 mmol), benzaldehyde (0.106 g, 1 mmol), two drops of piperidine, 1 drop of acetic acid and activated silica gel (1 g) into a dry mortar.
2) Mix the contents well with pestle add 4-5 drops of dichloromethane and transfer it into into microwave flask (beaker, 50 ml).
3) Place the flask/beaker into the Microwave oven and switch on the magnetic stirrer.
   - In case of Microwave synthesizer, attach the reflux condenser to the flask.
   - Cover the flask with a petridish loaded with ice cubes (provides condensing effect), alternatively funnel can be used. Beaker containing cold water can be used as a heating sink.
4) Irradiate the reaction mixture present in the beaker with Microwaves at 100% intensity for 15 minutes, with intermittent cooling (to avoid charring) for 1 minute after every 2 minutes of microwave irradiation.

5) During cooling stir the reaction mixture with spatula.
6) Turn off the power supply to the Microwave oven.
7) After few minutes, remove the beaker from the Microwave oven.
8) Cool and extract the reaction mixture with methanol (3 x 10 ml) and evaporate the solvent.
9) Purify the product by recrystallization with aqueous ethanol (70%).

Comparison of conventional and microwave method

	Conventional	***Microwave***
Reaction time	15 h	15 min
Reaction temperature	110 °C	100% intensity (850 W)
Colour	Light brown crystals	
Melting point	> 300 °C	
Percentage yield	80%	90%

METHOD 2 ***(using toluene as solvent)***

Chemicals required

- Thiazolidine-2,4-dione
- Benzaldehyde
- Piperidine
- Acetic acid
- 4-5 molecular sieves
- Toluene

Procedure:

1) Load toluene (10 ml), thiazolidine-2,4-dione (0.117 g, 1 mmol), benzaldehyde (0.106 g, 1 mmol), two drops of piperidine, 1 drop of acetic acid and 4-5 molecular sieves into microwave flask (beaker, 50 ml).
2) Place the flask/beaker into the Microwave oven and switch on the magnetic stirrer.
    - In case of Microwave synthesizer, attach the reflux condenser to the flask.
    - Cover the flask with a petridish loaded with ice cubes (provides condensing effect), alternatively funnel can be used. Beaker containing cold water can be used as a heating sink.

3) Irradiate the reaction mixture present in the beaker with Microwaves at 100% intensity for 25 minutes, with intermittent cooling (to avoid charring) for 1 minute after every 5 minutes of microwave irradiation.
4) During cooling stir the reaction mixture with spatula.
5) Turn off the power supply to the Microwave oven.
6) After few minutes, remove the beaker from the Microwave oven.
7) Cool and extract the reaction mixture with methanol (3 x 10 ml) and evaporate the solvent.
8) Purify the product by recrystallization with aqueous ethanol (70%).

Note: Monitor the progress of reaction through thin-layer chromatography (TLC).

Comparison of conventional and microwave method

	Conventional	***Microwave***
Reaction time	15 h	25 min
Reaction temperature	110 °C	100% intensity (850 W)
Colour	Light brown crystals	
Melting point	> 300 °C	
Percentage yield	80%	93%

Post Lab Questions:

1) Outline the Knoevenagel condensation reaction.
2) What is azeotropic condition?
3) What are the different catalysts useful in the Knoevenagel condensation?

HANTZSCH 1, 4-DIHYDROPYRIDINE SYNTHESIS

The multicomponent condensation of aryl/heteroaryl aldehydes, β-ketoester and ammonia in one-pot forms 4-aryl/heteroaryl-1,4-dihydropyridine (1,4-DHP). This reaction is known as Hantzsch 1,4-dihydropyridine synthesis. This reaction is catalyzed by mineral acids and metal salts. Variety of other catalysts can be used.

Ar—CHO + β-Ketoester + CH_3COONH_4 → (H^+, C_2H_5OH) 4-Aryl/heteroaryl-1,4-dihydropyridine

Aryl/heteroaryl aldehyde

β-Ketoester

Ammonium acetate

4-Aryl/heteroaryl-1,4-dihydropyridine

Reaction Mechanism

Aryl aldehyde

CH_3COONH_4

NH_3

β-Keto ester (enol form)

Ethyl acetoacetate (enol form)

Knoevenegal condensation

Aryledene product

Amino ester

$-NH_3$

$-H_2O$

4-Aryl/heteroaryl-1,4-dihydropyridine

EXPERIMENT 27

Preparation of 1,4-Dihydropyridine

Principle: The multicomponent condensation of benzaldehyde, ethyl acetoacetate and ammonia under produces diethyl-2,6-dimethyl-4-phenyl-1,4-dihydropyridine-3,5-dicarboxylate. This is example for Hanzsch 1,4-dihydropyridine synthesis.

Benzaldehyde + 2 Ethyl acetoacetate + NH_3 (Ammonia) $\xrightarrow[\text{12 min}]{\text{MW 850 W}}$ Diethyl-2,6-dimethyl-4-phenyl-1,4-dihydropyridine-3,5-dicarboxylate

Chemicals required

- Benzaldehyde
- Ethyl acetoacetate
- Aqueous ammonia (25%)
- Ethanol

Procedure:

1) Load benzaldehyde (0.212 g, 2 mmol), ethyl acetoacetate (0.52 g, 4 mmol), aqueous ammonia (25%, 3 ml) and ethanol (10 ml) into microwave flask (beaker, 50 ml).
2) Place the flask/beaker into the Microwave oven and switch on the magnetic stirrer.
   - In case of Microwave synthesizer, attach the reflux condenser to the flask.
   - Cover the flask with a petridish loaded with ice cubes (provides condensing effect), alternatively funnel can be used. Beaker containing cold water can be used as a heating sink.

3) Irradiate the reaction mixture present in reaction flask with Microwaves at 100% intensity for 12 minutes, with intermittent cooling (to avoid charring) for 1 minute after every 3 minutes of microwave irradiation.
4) Turn off the power supply to the Microwave oven.
5) After few minutes, remove the flask from the Microwave oven.
6) Cool and allow the reaction mixture to solidify.
7) Filter to isolate the solid crystals.
8) Purify the solid product by recrystallization with aqueous methanol.

Note: Monitor the progress of reaction through thin-layer chromatography (TLC).

Comparison of conventional and microwave method

	Conventional	*Microwave*
Reaction time	6 h	12 min
Reaction temperature	70-80 °C	100% intensity (850 W)
Colour	Light yellow	
Melting point	160-162 °C	
Percentage yield	78%	78%

Post Lab Questions:

1) List the 1, 4-dihydropyridine containing therapeutics and mention their biological importance.
2) What is Hantzsch 1.4-dihydropyridine synthesis?
3) What are the different catalysts suitable for the Hantsch 1, 4-dihydropyridine synthesis?

EXPERIMENT 28

Preparation of N-Aryl-1,4-dihydropyridine by (Solid Phase Reaction)

Principle: The multicomponent condensation of furfuraldehyde, ethyl aceto acetate and 4-amino phenol in the presence of *p*-toluene sulphonic acid (*p*-TSA) under solid phase reaction condition (silica gel) produces diethyl-4(furan-2-yl)-2,6-dimethyl-1-phenyl-1,4-dihydropyridine-3,5-dicarboxylate-1,4-dihydropyridine (1,4-DHP). This is example for Hantsch 1,4-dihydropyridine synthesis. Microwave irradiation, solventless and solid phase improves the product yields. Silica gel adsorbs the water molecules (adsorbent).

Ethyl acetoacetate + Furfuraldehyde + 4-Aminophenol + Ethyl acetoacetate

MW 850 W, 12 min, pTSA

Diethyl-4-(furan-2-yl)-2,6-dimethyl-1-phenyl-1,4dihydropyridine-3,5-dicarboxylate

Chemicals required

- Furfuraldehyde
- Ethyl acetoacetate
- 4-Aminophenol
- Silica gel (activated)
- -Toluene sulphonic acid (*p*TSA)
- Methanol (80%)

Note: Monitor the progress of reaction through thin-layer chromatography (TLC).

Procedure:

1) Load furfuraldehyde (0.288 g, 3 mmol), ethyl acetoacetate (0.78 g, 6 mmol), 4-aminophenol (0.327 g, 3 mmol) and silica gel (3 g) into a mortar along with a pinch of *p*-toluene sulphonic acid.
2) Triturate the contents with pestle and transfer the reaction mixture into microwave flask (beaker, 50 ml).
3) Place the flask/beaker into the Microwave oven and switch on the magnetic stirrer.
   - In case of Microwave synthesizer, attach the reflux condenser to the flask.
   - Cover the flask with a petridish loaded with ice cubes (provides condensing effect), alternatively funnel can be used. Beaker containing cold water can be used as a heating sink.
4) Irradiate the reaction mixture present in the beaker with Microwaves at 100% intensity for 12 minutes, with intermittent cooling (to avoid charring) for 1 minute after every 2 minutes of microwave irradiation.
5) During cooling stir the reaction mixture with spatula.
6) Turn off the power supply to the Microwave oven.
7) After few minutes, remove the beaker from the Microwave oven.
8) After the microwave irradiation, cool and extract the reaction mixture with methanol.
9) Heat the methanol extract, filter in hot condition and allow it to cool.
10) Collect the solid crystals formed and purify the product by recrystallization with methanol (80%) in water.

Comparison of conventional and microwave method

	Conventional	*Microwave*
Reaction time	12 h	12 min
Reaction temperature	70-80 °C	100% intensity (850 W)
Colour	Light pink	
Melting point	67-69 °C	
Percentage yield	45%	68%

Post Lab Questions:

1) What is Hantsch-1, 4-dihydropyridine synthesis?
2) What are the different catalysts useful in Hantsch-1, 4-dihydropyridine synthesis?
3) Does microwave reactions need solvent as media?

EXPERIMENT 29

Preparation of N-Aryl-1,4-dihydropyridine from Schiffbase (Solid Phase Reaction)

Principle: The condensation of furfuraldehyde and 4-aminophenol generates furanylmethylidene amino phenol (Schiff base). This reaction can be effected suing microwave irradiation coupled with solid phase conditions (silica gel). Silica gel acts as an adsorbent and adsorbs the released water molecules.

Furfuraldehyde (furan–CHO) + 4-Aminophenol (H_2N–C_6H_4–OH) → (MW 765 W, 6 min, Silica Gel) Furanylmethylidene aminophenol (furan–CH=N–C_6H_4–OH)

The condensation of Schiff base and ethyl acetoacetate in microwave assisted reaction. The conventional one pot synthesis, produced aceto acetanilide as an intermediate. This is an reaction limiting factor. Hence, the alternative method of condensation between Schiff base and ethyl acetoacetate is adopted.

Ethyl acetoacetate + Furanylmethylidene aminophenol + Ethyl acetoacetate

MW 850 W, 15 min, pTSA

Diethyl-4-(furan-2-yl)-2,6-dimethyl-1-phenyl-1,4dihydropyridine-3,5-dicarboxylate

Chemicals required

- Schiff base
- Ethyl acetoacetate
- 4-Aminophenol
- Silica gel (activated)
- *p*-Toluene sulphonic acid (*p*TSA)
- Methanol (80%)

Procedure:

1) Load Schiff base (0.288 g, 1.6 mmol) and ethyl acetoacetate (0.78 g, 6 mmol) into a dry mortar along with a pinch of *p*-toluene sulphonic acid.
2) Triturate the contents with pestle and transfer the reaction mixture into microwave flask (beaker, 50 ml).
3) Place the flask/beaker into the Microwave oven and switch on the magnetic stirrer.
   - In case of Microwave synthesizer, attach the reflux condenser to the flask.
   - Cover the flask with a petridish loaded with ice cubes (provides condensing effect), alternatively

funnel can be used. Beaker containing cold water can be used as a heating sink.

4) Irradiate the reaction mixture present in the beaker with Microwaves at 100% intensity for 15 minutes, with intermittent cooling (to avoid charring) for 1 minute after every 2 minutes of microwave irradiation.
5) During cooling stir the reaction mixture with spatula.
6) Turn off the power supply to the Microwave oven.
7) After few minutes, remove the beaker from the Microwave oven.
8) After the microwave irradiation, cool and extract the reaction mixture with ethyl acetate.
9) Dry the extract over sodium sulphate and evaporate under vaccum
10) Collect the solid crystals formed and purify the product by recrystallization with methanol (80%) in water.

Note: Monitor the progress of reaction through thin-layer chromatography (TLC).

Pinch of silica gel can be used in this reaction. In that case, reaction vessel need to be stirred with spatula during intermittent cooling.

Comparison of conventional and microwave method

	Conventional	***Microwave***
Reaction time	12 h	15 min
Reaction temperature	70-80 °C	100% intensity (850 W)
Colour	Light pink	
Melting point	67-69 °C	
Percentage yield	45%	72%

Post Lab Questions:

1) What is Hantsch-1, 4-dihydropyridine synthesis?
2) What are the different catalysts useful in Hantsch -1, 4-dihydropyridine synthesis?

RADZISZEWSKI IMIDAZOLE SYNTHESIS

The condensation of α-dicarbonyl compound (diketone), aryl/heteroaryl aldehyde and ammonia/ammonium acetate (two equivalents) in ethanol medium forms 2,4,5-trisubstituted imidazole. Ammonium carbonate, aqueous ammonia and other amines (Aryl amine) can be used in place of ammonium acetate. The use of aryl amines and other amines produces 1,2,4,5-tetrasubstituted imidazoles. The triarylimidazoles are the important organic compounds of medicinal interest. This reaction is known as Radziszewski synthesis. It is very important synthetic strategy for the chemical construction of imidazole scaffolds. Imidazole scaffolds are reported to have several therapeutic potential.

Diketone + Ar—CHO (Aryl/heteroaryl aldehyde) + CH_3COONH_4 (Ammonium acetate) ⟶ 2,4,5-Trisubstituted imidazole

Reaction mechanism

CH_3COONH_4 → NH_3

Dicarbonyl compound

Arylaldehyde

Schiff base of dicarbonyl compound

Schiff base of arylaldehyde

$-H_2O$

2,4,5-Triaryl imidazole

EXPERIMENT 30

Preparation of 2,4,5-Triaryl Imidazole (Solventless Reaction)

Principle: The multicomponent condensation of benzil (1,2-diketone), benzaldehyde and ammonium acetate forms 2,4,5-triphenylimidazole. The conventional methods utilize different catalysts. The microwave method under solid phase (silica gel) reaction conditions produces higher yield (efficient method). The solvent-free method is considered as green chemistry approach. The silica gel adsorbs water molecules released during the course of reaction.

CH_3COONH_4 Ammonium acetate
Benzil + OHC– Benzaldehyde
CH_3COONH_4 Ammonium acetate
MW 850 W, 12 min
Silica gel
2,4,5-Triphenyl imidazole

Chemicals required

- Benzil
- Benzaldehyde
- Ammonium acetate
- Silica gel (activated)

Procedure:

1) Load benzil (0.420 g, 2 mmol), benzaldehyde (0.272 g, 2.6 mmol), ammonium acetate (1.48 g, 19.2 mmol) and silica gel (2 g) into a mortar.
2) Grind the mixture with the help of pestle and transfer it into microwave flask (beaker, 50 ml).
3) Place the flask/beaker into the Microwave oven and switch on the magnetic stirrer.
    - In case of Microwave synthesizer, attach the reflux condenser to the flask.

- Cover the flask with a petridish loaded with ice cubes (provides condensing effect), alternatively funnel can be used. Beaker containing cold water can be used as a heating sink.

4) Irradiate the reaction mixture present in the beaker with Microwaves at 100% intensity for 12 minutes, with intermittent cooling (to avoid charring) for 1 minute after every 2 minutes of microwave irradiation.
5) During cooling stir the reaction mixture with spatula.
6) Turn off the power supply to the Microwave oven.
7) After few minutes, remove the beaker from the Microwave oven.
8) Cool and extract the reaction mixture with methanol.
9) Heat the methanol extract, filter in hot condition and allow it to cool.
10) Filter the solid precipitate and purify the crude product by recrystallization with ethanol.
11) In case of no precipitate, evaporate the solvent to get the product.

Note: Monitor the progress of reaction through thin-layer chromatography (TLC).

Comparison of conventional and microwave method

	Conventional	***Microwave***
Reaction time	9 h	12 min
Reaction temperature	190-200 °C	100% (850 W)
Colour	White	
Melting point	271-273 °C	
Percentage yield	80%	92%

EXPERIMENT 31

Preparation of 2,4,5-Triaryl Imidazole (using glacial acetic acid as a solvent)

Principle: The multicomponent condensation of benzil (1,2-diketone), benzaldehyde and ammonium acetate forms 2,4,5-triphenylimidazole. The conventional methods utilize different catalysts. The microwave method under solid phase (silica gel) reaction conditions produces higher yield (efficient method). The solvent-free method is considered as green chemistry approach. The silica gel adsorbs water molecules released during the course of reaction.

CH_3COONH_4 Ammonium acetate + Benzil + OHC–Benzaldehyde + CH_3COONH_4 Ammonium acetate → (MW 850 W, 10 min, Gla CH_3COOH) 2,4,5-Triphenyl imidazole

Chemicals required

- Benzil
- Benzaldehyde
- Ammonium acetate
- Glacial acetic acid

Procedure:

1) Load benzil (0.420 g, 2 mmol), benzaldehyde (0.272 g, 2.6 mmol), ammonium acetate (1.48 g, 19.2 mmol) and glacial acetic acid (5 ml) into microwave flask (beaker, 50 ml).
2) Place the flask/beaker into the Microwave oven and switch on the magnetic stirrer.
   - In case of Microwave synthesizer, attach the reflux condenser to the flask.
   - Cover the flask with a petridish loaded with ice cubes (provides condensing effect), alternatively funnel can be used. Beaker containing cold water can be used as a heating sink.

3) Irradiate the reaction mixture present in the beaker with Microwaves at 100% intensity for 12 minutes, with intermittent cooling (to avoid charring) for 1 minute after every 2 minutes of microwave irradiation.
4) Turn off the power supply to the Microwave oven.
5) After few minutes, remove the beaker from the Microwave oven.
6) Cool and extract the reaction mixture with methanol.
7) Heat the methanol extract, filter in hot condition and allow it to cool.
8) Filter the solid precipitate and purify the crude product by recrystallization with ethanol.
9) In case of no precipitate, evaporate the solvent to get the product.

Note: Monitor the progress of reaction through thin-layer chromatography (TLC).

Comparison of conventional and microwave method

	Conventional	***Microwave***
Reaction time	9 h	10 min
Reaction temperature	190-200 °C	100% (850 W)
Colour	White	
Melting point	271-273 °C	
Percentage yield	80%	92%

Post Lab Questions:

1) Why activated silica is used in this reaction?
2) What are the alternatives to silica gel?
3) List the imidazole containing therapeutics and mention their biological importance.
4) What is the role of ammonium acetate? What are the alternatives for ammonium acetate?

EXPERIMENT 32

Preparation of Tetraaryl Imidazole

Principle: The multicomponent condensation of benzil (1,2-diketone), anisaldehyde, ammonium acetate and 4-aminophenol forms 4-(2,4,5-triphenyl-1H-imidazol-1-yl)-phenol. The conventional methods utilize different catalysts. The microwave method under solid phase (silica gel) reaction conditions produces higher yield (efficient method). The solvent-free method is considered as green chemistry approach. The silica gel adsorbs water molecules released during the course of reaction. In this reaction triarylimidazole is minor product and tetraarylimidazole is major product.

CH_3COONH_4 Ammonium acetate
Benzil
4-Aminophenol
Anisaldehyde
MW 850 W, 12 min
Gla CH_3COOH
Tetraphenyl imidazole

Chemicals required

- Benzil
- Anisaldehyde
- 4-Aminophenol
- Ammonium acetate
- Silica gel (activated)
- Methanol

Procedure:

1) Load benzil (0.420 g, 2 mmol), anisaldehyde (0.272 g, 2 mmol), 4-aminophenol (0.218 g, 2 mmol), ammonium acetate (1.48 g, 19.2 mmol) and silica gel (2 g) into a mortar.
2) Grind the mixture with the help of pestle and transfer it into microwave flask (beaker, 50 ml).

3) Place the flask/beaker into the Microwave oven and switch on the magnetic stirrer.
   - In case of Microwave synthesizer, attach the reflux condenser to the flask.
   - Cover the flask with a petridish loaded with ice cubes (provides condensing effect), alternatively funnel can be used. Beaker containing cold water can be used as a heating sink.
4) Irradiate the reaction mixture present in the beaker with Microwaves at 100% intensity for 12 minutes, with intermittent cooling (to avoid charring) for 1 minute after every 2 minutes of microwave irradiation.
5) During cooling stir the reaction mixture with spatula.
6) Turn off the power supply to the Microwave oven.
7) After few minutes, remove the beaker from the Microwave oven.
8) Cool and extract the reaction mixture with methanol.
9) Heat the methanol extract, filter in hot condition and allow it to cool.
10) Collect the solid crystals and purify the crude product by recrystallization with ethanol.

Note: Monitor the progress of reaction through thin-layer chromatography (TLC).

Comparison of conventional and microwave method

	Conventional	***Microwave***
Reaction time	14 h	12 min
Reaction temperature	190-200 °C	100% intensity (850 W)
Colour	White	
Melting point	243-245 °C	
Percentage yield	70%	80%

Optional Activity: The reaction can be performed using different aldehydes and amines.

EXPERIMENT 33

Preparation of Tetraphenyl Imidazole via Schiff base (Solventless Reaction)

Principle: This method is developed to prepare tetraphenylimidaozle selectively. The multicomponent condensation of benzil (1,2-diketone), Schiff base (4-{[(4-methoxyphenyl)methylidene] amino}phenol) and ammonium acetate forms 4-(2,4, 5-triphenyl-1H-imidazol-1-yl)-phenol. Schiff base (4-{[(4-methoxyphenyl) methyl-idene]amino}phenol) can be prepared by the microwave condensation of 4-methoxybenzaldehyde and 4-aminophenol.

Anisaldehyde + 4-Aminophenol → (MW 850 W, 12 min) → 4-{[(4-Methoxyphenyl)methylidene] amino}phenol (Schiff base)

Benzil + 4-{[(4-Methoxyphenyl)methylidene] amino}phenol (Schiff base) → (MW 850 W, 12 min; Ammonium acetate; Gla Acetic acid) → Tetraphenyl imidazole

The conventional methods utilize different catalysts. The microwave method under solid phase (silica gel) reaction conditions produces higher yield (efficient method). The solvent-free method is considered as green chemistry approach. The silica gel adsorbs water molecules released during the course of reaction.

Chemicals required

- Benzil
- Schiff base (4-{[(4-methoxyphenyl) methylidene] amino}phenol)
- Ammonium acetate
- Silica gel (activated)
- Methanol

Procedure:

1) Load benzil (0.420 g, 2 mmol), Schiff base (4-{[(4-methoxyphenyl)methylidene] amino}phenol) (0.218 g, 1 mmol), ammonium acetate (1.48 g, 19.2 mmol) and silica gel (2 g) into a mortar.
2) Grind the mixture with the help of pestle and transfer it into microwave flask (beaker, 50 ml).
3) Place the flask/beaker into the Microwave oven and switch on the magnetic stirrer.
   - In case of Microwave synthesizer, attach the reflux condenser to the flask.
   - Cover the flask with a petridish loaded with ice cubes (provides condensing effect), alternatively funnel can be used. Beaker containing cold water can be used as a heating sink.
4) Irradiate the reaction mixture present in the beaker with Microwaves at 100% intensity for 12 minutes, with intermittent cooling (to avoid charring) for 1 minute after every 2 minutes of microwave irradiation.
5) During cooling stir the reaction mixture with spatula.
6) Turn off the power supply to the Microwave oven.
7) After few minutes, remove the beaker from the Microwave oven.
8) Cool and extract the reaction mixture with methanol.
9) Heat the methanol extract, filter in hot condition and allow it to cool.
10) Collect the solid crystals and purify the crude product by recrystallization with ethanol.

Note: Monitor the progress of reaction through thin-layer chromatography (TLC).

Comparison of conventional and microwave method

	Conventional	*Microwave*
Reaction time	14 h	12 min
Reaction temperature	190-200 °C	100% intensity (850 W)
Colour	White	
Melting point	243-245 °C	
Percentage yield	70%	80%

EXPERIMENT 34

Preparation of Tetraphenyl Imidazole via Schiff base (Solvent Reaction)

Principle: This method is developed to prepare tetraphenylimidaozle selectively. The multicomponent condensation of benzil (1,2-diketone), Schiff base (4-{[(4-methoxyphenyl)methylidene] amino}phenol) and ammonium acetate forms 4-(2,4,5-triphenyl-1H-imidazol-1-yl)-phenol. Schiff base (4-{[(4-methoxyphenyl) methylidene] amino}phenol) can be prepared by the microwave condensation of 4-methoxybenzaldehyde and 4-aminophenol.

Anisaldehyde + 4-Aminophenol → (MW 850 W, 12 min) 4-{[(4-Methoxyphenyl)methylidene] amino}phenol (Schiff base)

Benzil + 4-{[(4-Methoxyphenyl)methylidene] amino}phenol (Schiff base) → MW 850 W, 12 min; Ammonium acetate; Gla Acetic acid → Tetraphenyl imidazole

The conventional methods utilize different catalysts. The microwave method under solid phase (silica gel) reaction conditions produces higher yield (efficient method). The solvent-free method is considered as green chemistry approach. The silica gel adsorbs water molecules released during the course of reaction.

Chemicals required

- Benzil
- Schiff base (4-{[(4-methoxyphenyl)methylidene] amino}phenol)
- Ammonium acetate
- Glacial acetic acid
- Methanol

Procedure:

1) Load benzil (0.420 g, 2 mmol), Schiff base (4-{[(4-methoxyphenyl)methylidene] amino}phenol) (0.218 g, 1 mmol), ammonium acetate (1.48 g, 19.2 mmol) and glacial acetic acid (5 ml) into microwave flask (beaker, 50 ml).
2) Place the flask/beaker into the Microwave oven and switch on the magnetic stirrer.
   - In case of Microwave synthesizer, attach the reflux condenser to the flask.
   - Cover the flask with a petridish loaded with ice cubes (provides condensing effect), alternatively funnel can be used. Beaker containing cold water can be used as a heating sink.
3) Irradiate the reaction mixture present in the beaker with Microwaves at 100% intensity for

12 minutes, with intermittent cooling (to avoid charring) for 1 minute after every 2 minutes of microwave irradiation.

4) Turn off the power supply to the Microwave oven.
5) After few minutes, remove the beaker from the Microwave oven.
6) Cool and extract the reaction mixture with methanol.
7) Heat the methanol extract, filter in hot condition and allow it to cool.
8) Collect the solid crystals and purify the crude product by recrystallization with ethanol.

Note: Monitor the progress of reaction through thin-layer chromatography (TLC).

Comparison of conventional and microwave method

	Conventional	***Microwave***
Reaction time	14 h	12 min
Reaction temperature	190-200 °C	100% intensity (850 W)
Colour	White	
Melting point	243-245 °C	
Percentage yield	70%	80%

MANNICH REACTION

Mannich condensation is one of the reactions of synthetic importance in organic chemistry. The condesation of an enolizable carbonyl compound, a secondary amine and formadehyde under acid catalysed reaction forms amino ketone product. This reaction is known as Mannich reaction. It involves the reaction between compounds possessing active hydrogen atoms (ketones, acids), formaldehyde and ammonia or primary/ secondary amines. Several Mannich bases are reported to possess antimicrobial, antitubercular and antiviral and other biological activities.

Carbonyl group (enolizable) + Formaldehyde + Secondary amine $\xrightarrow{HCl}$ Amino ketone

EXPERIMENT 35

Preparation of Mannich Base

Principle In the present experiment, a mixture of acetophenone, paraformaldehyde and diethylamine, in the presence concentrated hydrochloric acid was irradiated with microwaves in a Borosil beaker. After completion of the reaction, reaction mixture was cooled and a small amount of acetone was added to induce precipitation. Under microwaves, reaction rate has improved considerably with an increase in overall yield and purity of the products.

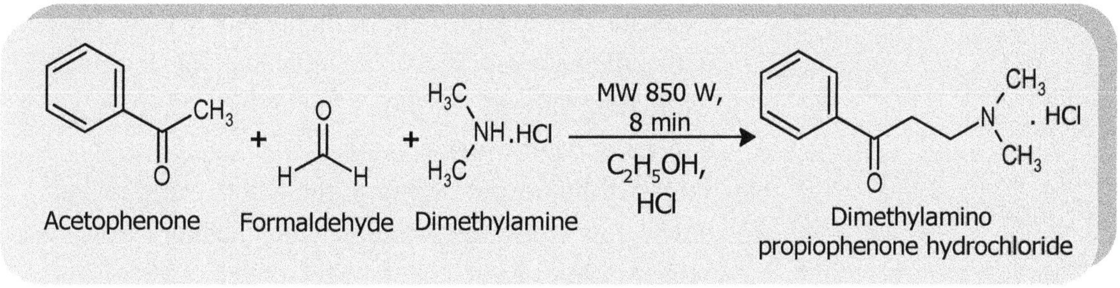

Acetophenone + Formaldehyde + Dimethylamine ($NH.HCl$) → (MW 850 W, 8 min; C_2H_5OH, HCl) Dimethylamino propiophenone hydrochloride

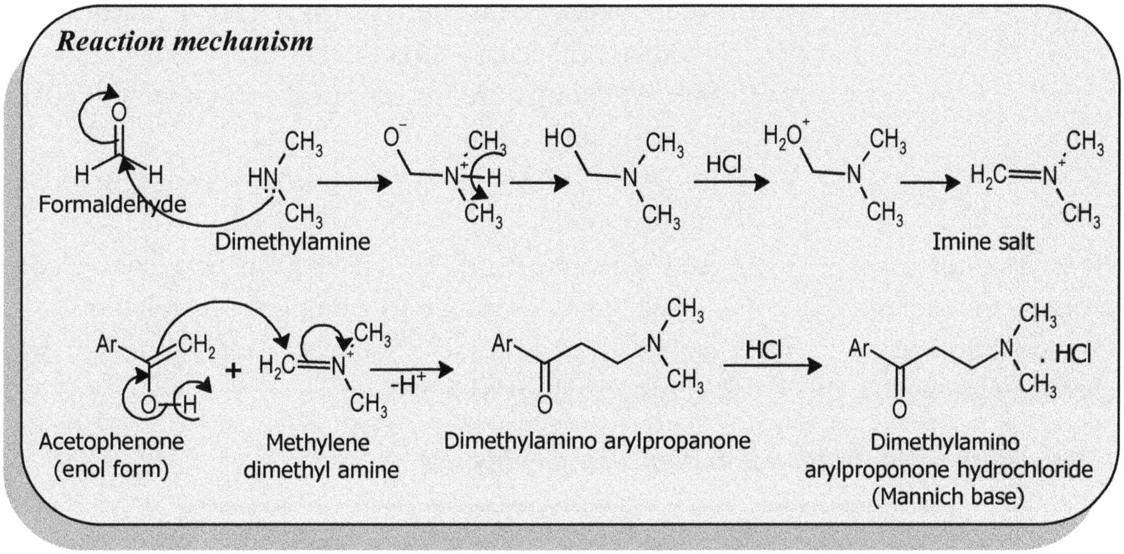

Reaction mechanism

Formaldehyde + Dimethylamine → → → (HCl) → Imine salt

Acetophenone (enol form) + Methylene dimethyl amine → ($-H^+$) Dimethylamino arylpropanone → (HCl) Dimethylamino arylproponone hydrochloride (Mannich base)

Chemicals required

- Acetophenone
- Dimethylamine
- Formaldehyde
- Ethanol
- Acetone

Procedure:

1) Load dimethyl amine (0.45 g, 10 mmol), formaldehyde (0.36 g, 12 mmol) and acetophenone (1.2 g, 10 mmol) into microwave flask (beaker, 50 ml).
2) Place the flask into the Microwave oven and switch on the magnetic stirrer.
   - In case of Microwave synthesizer, attach the reflux condenser to the flask.
   - Cover the flask with a petridish loaded with ice cubes (provides condensing effect), alternatively funnel can be used. Beaker containing cold water can be used as a heating sink.

3) Irradiate the reaction mixture present in the beaker with Microwaves at 100% intensity for 8 minutes, with intermittent cooling (to avoid charring) for 2 minutes after every minute.
4) Turn off the power supply to the Microwave oven.
5) After few minutes, remove the beaker from the Microwave oven.
6) Add absolute ethanol (5 ml) and concentrated hydrochloric acid (2 ml) to it.
7) Subject the mixture for microwave irradiation at 60% intensity for 6 minutes.
8) Remove the beaker from the oven and cool it to room temperature.
9) Add acetone (5 ml) to it and keep in a cool place (refrigerator) overnight for complete precipitation.
10) Isolate the crude product and purify by recrystallization with ethanol

Note: Monitor the progress of reaction through thin-layer chromatography (TLC).

Comparison of conventional and microwave method

	Conventional	***Microwave***
Reaction time	15 h	8 min; 6 min
Reaction temperature	70-80 °C	100%; 60%
Colour	White	
Melting point	153-154 °C	
Percentage yield	76%	74%

Post Lab Questions:

1) What kinds are amines are used to perform Mannich reactions?

PHILIPS REACTION

The o-Phenylenediamines condense with aliphatic organic acids to produce benzimidazoles. This reaction is catalyzed by mineral acid (eg: hydrochloric acid). Aromatic acids produce products only in sealed reaction tube. Aromatic aldehydes also can be condensed with the *o*-phenylenediamine to produce 2-aryl substituted imidazoles.

o-Phenylenediamine + H-COOH (Formic acid) $\xrightarrow{HCl}$ Benzimidazole

Reaction mechanism

o-Phenylenediamine + H-COOH → intermediate $\xrightarrow{-H_2O}$ intermediate $\xrightarrow{-H2O}$ Benzimidazole

EXPERIMENT 36

Preparation of Benzimidazole

Principle: Benzimidazole can be prepared by heating *o*-phenylenediamine with formic acid. This reaction involves simple condensation with elimination of 2 molecules of water followed by cyclization under acidic conditions.

o-Phenylenediamine + H-COOH (Formic acid) $\xrightarrow{\text{MW 510 W, 8 min}}$ Benzimidazole

Chemicals required

- *o*-phenylenediamine
- Formic acid
- Sodium hydroxide (10%) solution

Note: Monitor the progress of reaction through thin-layer chromatography (TLC).

Procedure:

1) Load *o*-phenylenediamine (0.648 g, 6 mmol) and formic acid (90%, 8 ml) into microwave flask (beaker, 50 ml) and mix well.
2) Place the beaker into the Microwave oven and switch on the magnetic stirrer.
   - In case of Microwave synthesizer, attach the reflux condenser to the flask.
   - Cover the flask with a petridish loaded with ice cubes (provides condensing effect), alternatively funnel can be used. Beaker containing cold water can be used as a heating sink.
3) Irradiate the reaction mixture present in beaker with Microwaves at 60% intensity for 8 minute, with intermittent cooling (to avoid charring) for 1 minute after every 4 minutes.
4) Turn off the power supply to the Microwave oven.
5) After few minutes, remove the beaker from the Microwave oven.
6) Alkalify the mixture with sodium hydroxide (10%) and isolate the crude benzimidazole by filtration.
7) Wash the crude benzimidazole with ice cold water.
8) Purify the crude benzimidazole recrystallization with boiling water.

Comparison of conventional and microwave method

	Conventional	***Microwave***
Reaction time	2 h	8 min
Reaction temperature	190-200 °C	60% intensity (510 W)
Colour	White	
Melting point	172-173 °C	
Percentage yield	90%	96%

Post Lab Questions:

1) What is Philips reaction?
2) What are the different catalytic conditions applicable for the Philips reaction?

EXPERIMENT 37

Preparation of 2-Phenylbenzimidazole

Principle: The condensation of *o*-phenylenediamine with arylaldehydes forms 2-phenylbenzimidazole. The reaction is catalyzed by acidic conditions. The catalyst *p*-toluenesulphonic acid and the acidic solvent dimethyl acetamide are useful. It is a condensation reaction involving elimination of water and ammonia. The *p*-toluenesulphonic acid is convenient under microwave conditions compared to hydrochloric acid.

o-Phenylene diamine + Benzaldehyde → (MW 850 W, 9 min; DMAC, PTSA) → 2-Phenylbenzimidazole

Chemicals required

- Benzaldehyde
- *o*-phenylenediamine
- *p*-Toluenesulphonic acid
- Dimethyl acetamide (DMAC)

Procedure:

1) Load benzaldehyde (0.16 g, 1.5 mmol), o-phenylene-diamine (0.108 g, 1 mmol) *p*-toluenesulphonic acid (0.02 g) and dimethyl acetamide (DMAC, 3 ml) into microwave flask (beaker, 50 ml) and mix well.
2) Place the flask/beaker into the Microwave oven and switch on the magnetic stirrer.

- In case of Microwave synthesizer, attach the reflux condenser to the flask.
- Cover the flask with a petridish loaded with ice cubes (provides condensing effect), alternatively funnel can be used. Beaker containing cold water can be used as a heating sink.

3) Irradiate the reaction mixture present in the beaker with Microwaves (100%; 9 min), with intermittent cooling for 2 minutes after every minute of microwave irradiation.
4) Turn off the power supply to the Microwave oven.
5) After few minutes, remove the flask from the Microwave oven.
6) Cool the mixture and add to a beaker containing cold water.
7) Collect the solid precipitate formed and recrystallize with aqueous methanol (80%).

Note: Monitor the progress of reaction through thin-layer chromatography (TLC).

Safety precaution: *o*-phenylenediamine is toxic and can be absorbed through the skin.

Comparison of conventional and microwave method

	Conventional	***Microwave***
Reaction time	4 h	9 min
Reaction temperature	90 °C	100% intensity (850 W)
Colour	White	
Melting point	290-292 °C	
Percentage yield	75%	83%

Post Lab Questions:

1) What is the role of dimethyl acetamide in this reaction?
2) List the benzimidazole compounds with biological importance.
3) What is the role of *p*-toluenesulphonic acid? What are the alternatives for *p*TSA?

EXPERIMENT 38

Preparation of Benzimidazol-2-one

Principle: The condensation of *o*-phenylenediamine with urea forms benzimidazol-2-one. This cyclization reaction is typical example for dry media reaction (green chemistry approach). Microwave method is more efficient compared to the conventional method. Urea is used in excess quantity, which catalyzes the reaction.

NH_2 NH_2 + H_2N (C=O) NH_2 —MW 850 W, 10 min→ Benzimidazol-2-one

o-Phenylene diamine   Urea   Benzimidazol-2-one

Chemicals required

- *o*-Phenylenediamine
- Urea
- Ethyl acetate

Procedure:

1) Load o-phenylenediamine (0.54 g, 5 mmol) and urea (0.3 g, 5 mmol) into a mortar.
2) Triturate the mixture well and transfer into microwave flask (beaker, 50 ml).

3) Place the flask/beaker into the Microwave oven and switch on the magnetic stirrer.
   - In case of Microwave synthesizer, attach the reflux condenser to the flask.
   - Cover the flask with a petridish loaded with ice cubes (provides condensing effect), alternatively funnel can be used. Beaker containing cold water can be used as a heating sink.
4) Irradiate the reaction mixture present in the beaker with Microwaves (100%; 10 min), with intermittent cooling for 2 minutes after every minute of microwave irradiation.
5) During cooling stir the reaction mixture with spatula.
6) Turn off the power supply to the Microwave oven.
7) After few minutes, remove the beaker from the Microwave oven.
8) Cool the reaction mixture and add to the beaker containing cold water.
9) Collect the crude solid and purify by recrystallization with ethyl acetate.

Note: Monitor the progress of reaction through thin-layer chromatography (TLC).

Safety precautions: *o*-phenylenediamine is toxic and can be absorbed through the skin.

Comparison of conventional and microwave method

	Conventional	*Microwave*
Reaction time	5 h	10 min
Reaction temperature	100 °C	100% intensity (850 W)
Colour	White	
Melting point	308-310 °C	
Percentage yield	74%	80%

Post Lab Questions:

1) What is dry media reaction?
2) What is the role of urea in this reaction?

EXPERIMENT 39

Preparation of Benzimidazolesulphonic Acid

Principle: The condensation of *o*-phenylenediamine with thiourea forms benzimidazol-2-thione. This cyclization reaction is typical example for dry media reaction (green chemistry approach). Microwave method is more efficient compared to the conventional method. Thiourea is used in excess quantity, which catalyzes the reaction. The benzimidazole-2-thione upon oxidation with alkaline potassium permanganate forms benzimidazole sulphonic acid. The enol form (keto-enol) of benzimidazole-2-thione (2-mercapto benzimidazole) undergoes permanganate oxidation and forms benzimidazolesulphonic acid.

Step 1: *Preparation of benzimidazol-2-thione*

Chemicals required

- *o*-Phenylenediamine
- Thiourea
- Ethyl acetate

Procedure:

1) Load o-phenylenediamine (0.54 g, 5 mmol) and thiourea (0.382 g, 6.36 mmol) into a mortar.
2) Triturate the mixture well and transfer into microwave flask (beaker, 50 ml).
3) Place the flask into the Microwave oven and switch on the magnetic stirrer.
    - In case of Microwave synthesizer, attach the reflux condenser to the flask.
    - Cover the flask with a petridish loaded with ice cubes (provides condensing effect), alternatively funnel can be used. Beaker containing cold water can be used as a heating sink.
4) Irradiate the reaction mixture present in the beaker with Microwaves (80%; 8 min), with intermittent cooling for 1 minute after every 2 minutes of microwave irradiation.
5) During cooling stir the reaction mixture with spatula.
6) Turn off the power supply to the Microwave oven.
7) After few minutes, remove the beaker from the Microwave oven.
8) Cool the reaction mixture and add to the beaker containing cold water.
9) Collect the crude solid and purify by recrystallization with aqueous ethanol.

Note: Monitor the progress of reaction through thin-layer chromatography (TLC).

Safety precautions: *o*-phenylenediamine is toxic and can be absorbed through the skin.

Comparison of conventional and microwave method

	Conventional	*Microwave*
Reaction time	4 h	8 min
Reaction temperature	100 °C	80% intensity (680 W)
Colour	White	
Melting point	293-295 °C	
Percentage yield	74%	74%

Step 2: ***Preparation of benzimidazole sulphonic acid***

Chemicals required

- Benzimidazole-2-thione
- Potassium permanganate
- Sodium hydroxide (50%) solution

Procedure:

1) Load benzimidazole-2-thione (1 g, 6.7 mmol) into a beaker (50 ml) containing water (10 ml).
2) Add sodium hydroxide solution (50%, 5 ml) and potassium permanganate (3.2 g, 20 mmol) into microwave flask (beaker, 50 ml) and mix well.
3) Place the flask into the Microwave oven and switch on the magnetic stirrer.
   - In case of Microwave synthesizer, attach the reflux condenser to the flask.
   - Cover the flask with a petridish loaded with ice cubes (provides condensing effect), alternatively funnel can be used. Beaker containing cold water can be used as a heating sink.
4) Irradiate the reaction mixture present in the beaker with Microwaves (90%; 4 min), with intermittent cooling for one minute after every 2 minutes of microwave irradiation.
5) Turn off the power supply to the Microwave oven.
6) After few minutes, remove the beaker from the Microwave oven and cool the reaction mixture.

7) Pour the reaction mixture into the beaker containing cold dilute hydrochloric acid and filter the solid precipitate.
8) Add sodium metabisulphite to decolurise, if required.
9) Purify the crude product by recrystallization with water.

Note: Monitor the progress of reaction through thin-layer chromatography (TLC).

Comparison of conventional and microwave method

	Conventional	***Microwave***
Reaction time	3 h	4 min
Reaction temperature	100 °C	90% intensity (765 W)
Colour	White	
Melting point	325-330 °C	
Percentage yield	76%	85%

Post Lab Questions:

1) What is dry media reaction?
2) What is the role of potassium permanganate in this reaction?

AMIDE SYNTHESIS

Amides can be synthesized by heating carboxylic acid and amines at higher temperature (140 to 210 °C). At lower temperature, the reaction produces ammonium salt instead of amides (in case of acetic acid it forms ammonium acetate instead of acetamide).

$$R{-}NH_2 + R_1C(=O)NH_2 \xrightarrow[]{\Delta,\ >140\ ^\circ C} R_1C(=O)NH{-}R$$

Amine Carboxylic acid Amide

EXPERIMENT 40

Preparation of N-Butylbenzamide

Principle: The n-butylbenzamide can be prepared by reacting *n*-butylamine and benzoic acid using *para*-toluene sulphonic acid as a catalyst. The reaction involves nucleophilic acylic substitution. The reaction requires no solvent and is conducted in solid support (silica gel). This kind of reaction is known as dry media reaction. The grinded mixture upon microwave irradiation forms n-butylbenzamide. Coupling microwave irradiation with dry media conditions allowed reactions to occur on a preparative scale. This method overcomes the risk of high pressure and explosion. The product is very pure and requires no purification.

$$H_9C_4{-}NH_2 + C_6H_5C(=O)OH \xrightarrow[p\text{-TSA, Silica gel}]{\text{MW 850 W, 10 min}} C_6H_5C(=O)NH{-}C_4H_9$$

n-Butylamine Benzoic acid *n*-Butylbenzamide

Chemicals required

- n-Butylamine
- Benzoic acid
- *p*-Toluenesulphonic acid
- Silica gel
- Ether
- Sodium bicarbonate
- Sodium sulphate

Note: Monitor the progress of reaction through thin-layer chromatography (TLC).

Procedure:

1) Load *n*-butylamine (0.365 g, 5 mmol), benzoic acid (0.61 g, 5 mmol), *p*-toluenesulphonic acid (0.02 g, 0.11 mmol) and silica gel (0.3 g) into a mortar.
2) Grind the mixture with the help of pestle and transfer into microwave flask (beaker, 50 ml).
3) Place the beaker into the Microwave oven and switch on the magnetic stirrer.
   - In case of Microwave synthesizer, attach the reflux condenser to the flask.
   - Cover the flask with a petridish loaded with ice cubes (provides condensing effect), alternatively funnel can be used. Beaker containing cold water can be used as a heating sink.
4) Irradiate the reaction mixture present in the beaker with Microwaves (100%; 10 min), with intermittent cooling for one minute after every 2 minutes of microwave irradiation.
5) During cooling stir the reaction mixture with spatula.
6) Turn off the power supply to the Microwave oven.
7) After few minutes, remove the flask from the Microwave oven.
8) Transfer the mixture into a beaker, cool and add a mixture of solvent ether (20 ml) and distilled water (5ml).
9) Separate the ether layer and wash with a saturated solution of sodium bicarbonate, dry over sodium sulphate and evaporate the ether to get the product *n*-butylbenzamide.

Comparison of conventional and microwave method

	Conventional	*Microwave*
Reaction time	2 h	10 min
Reaction temperature	80-90 °C	100% intensity (850 W)
Colour	White	
Melting point	39-40 °C	
Percentage yield	86%	88%

Post Lab Questions:

1) How amides can be synthesised?
2) What is the role of *para*-toluenesulphonic acid in this reaction?

EXPERIMENT 41

Preparation of Phthalimide

Principle: Phthalimide is synthesised by the reaction between phthalic acid and ammonia. It forms uncyclized amide, which upon cyclization at higher temperature forms phthalimide.

Phthalic anhydride + NH_3 $\xrightarrow[-H_2O]{\text{MW 850 W, 4 min}}$ Phthalimide

Chemicals required

- Phthalic anhydride
- Concentrated ammonia solution

Procedure:

1) Mix phthalic anhydride (1.0 g, 6.7 mmol) and concentrated ammonia (5 ml) solution together into microwave flask (beaker, 50 ml).
2) Place the flask into the Microwave oven and switch on the magnetic stirrer.
   - In case of Microwave synthesizer, attach the reflux condenser to the flask.
   - Cover the flask with a petridish loaded with ice cubes (provides condensing effect), alternatively funnel can be used. Beaker containing cold water can be used as a heating sink.
3) Irradiate the reaction mixture present in reaction flask with Microwaves (100%; 4 min).
4) Turn off the power supply to the Microwave oven.
5) Remove the beaker from the oven.
6) The reaction mixture solidifies upon cooling and effloresces as a white powder.
7) Cool and collect the solid product, dry and weigh.
8) Recrystallize if necessary from alcohol.

Note: Monitor the progress of reaction through thin-layer chromatography (TLC).

Comparison of conventional and microwave method

	Conventional	*Microwave*
Reaction time	3 h	4 min
Reaction temperature	80-90 °C	100% intensity (850 W)
Colour	White	
Melting point	231-234 °C	
Percentage yield	80%	86%

Post Lab Questions:

1) Name other dehydrating agents used in laboratories?
2) Name fewer dicarboxylic acids that undergo this kind of reaction?

HALOFORM REACTION

Methyl ketones and secondary alcohols reacts with hypohalites ($-OX^-$) and forms trihalo methane (haloform). Acetone (methyl ketone) reacts with sodium hypoiodate and forms iodoform.

$$R-CO-CH_3 + NaOI \longrightarrow CHI_3 + R-CO-ONa$$

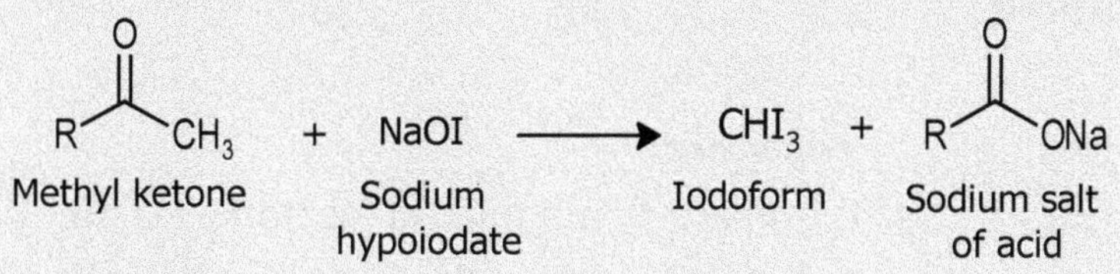

Methyl ketone + Sodium hypoiodate → Iodoform + Sodium salt of acid

Reaction mechanism

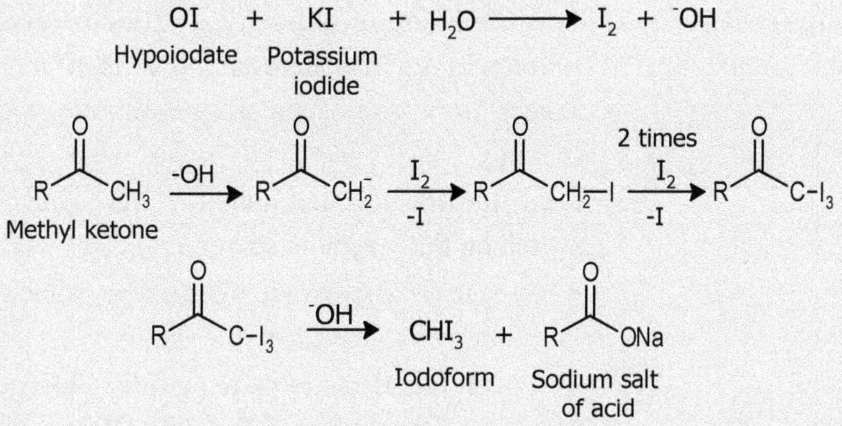

$$OI + KI + H_2O \longrightarrow I_2 + {}^-OH$$

Hypoiodate + Potassium iodide

$$R-CO-CH_3 \xrightarrow{-OH} R-CO-CH_2^- \xrightarrow[-I]{I_2} R-CO-CH_2-I \xrightarrow[-I]{I_2 \text{ (2 times)}} R-CO-C-I_3$$

Methyl ketone

$$R-CO-C-I_3 \xrightarrow{{}^-OH} CHI_3 + R-CO-ONa$$

Iodoform + Sodium salt of acid

EXPERIMENT 42

Preparation of Iodoform

Principle: Acetone (methyl ketone) reacts with sodium hypoiodate and forms iodoform.

$$CH_3COCH_3 + NaOI \xrightarrow{\text{MW 425 W, 4 min}} CHI_3 + CH_3COONa$$

Acetone + Sodium hypoiodate → Iodoform + Sodium acetate

Chemicals required

- Acetone
- Sodium hydroxide solution (10%)
- Iodine
- Potassium iodide
- Methanol (50%)

Procedure:

1) Prepare a mixture of acetone (1 ml, 1.36 mmol), in distilled water (10 ml) and sodium hydroxide solution (5 ml) microwave flask (beaker, 50 ml).
2) Add a solution of iodine (2 g of iodine dissolved in a solution of 4 g of potassium iodide in 10 ml of distilled water) drop wise to the above mixture with constant shaking.
3) Place the flask/beaker into the Microwave oven and switch on the magnetic stirrer.
   - In case of Microwave synthesizer, attach the reflux condenser to the flask.
   - Cover the flask with a petridish loaded with ice cubes (provides condensing effect), alternatively funnel can be used. Beaker containing cold water can be used as a heating sink.
4) Irradiate the reaction mixture present in reaction flask with Microwaves at 50% intensity for 4 minutes, with intermittent cooling for 2 min after every minute of microwave irradiation.
5) Turn off the power supply to the Microwave oven.
6) Cool the mixture and filter to remove the separated product.
7) Recrystallize with methanol (50%).

Note: Monitor the progress of reaction through thin-layer chromatography (TLC).

Comparison of conventional and microwave method

	Conventional	***Microwave***
Reaction time	1 h	4 min
Reaction temperature	60-70 °C	50% intensity (425 W)
Colour	Yellow	
Melting point	119 °C	
Percentage yield	76%	80%

Post Lab Questions:

1) What class of compounds undergoes haloform reactions?
2) What is the role of alkali in this reaction?

PAAL-KNORR PYRAZOLE SYNTHESIS

Paal-Knorr Pyrazole synthesis involves the condensation of 1,3-dicarbonyl compounds with hydrazines.

H_2N-NH_2 + 1,3-Dicarbonyl compound (R, R_1) → Pyrazole

Hydrazine 1,3-Dicarbonyl compound Pyrazole

EXPERIMENT 43

Preparation of 3-Methyl-1-phenyl pyrazole-5-one

Principle: The condensation of ethyl acetoacetate and phenylhydrazine forms 3-methyl-1-phenyl-pyrazole-5-one. This is typical example for the Paal-Knorr pyrazole synthesis.

Phenylhydrazine + Ethyl acetoacetate (OC_2H_5, CH_3) → (MW 850 W, 8 min, Ether) 3-Methyl-1-phenyl-pyrazole-5-one (CH_3)

Phenylhydrazine Ethyl acetoacetate 3-Methyl-1-phenyl-pyrazole-5-one

Reaction mechanism

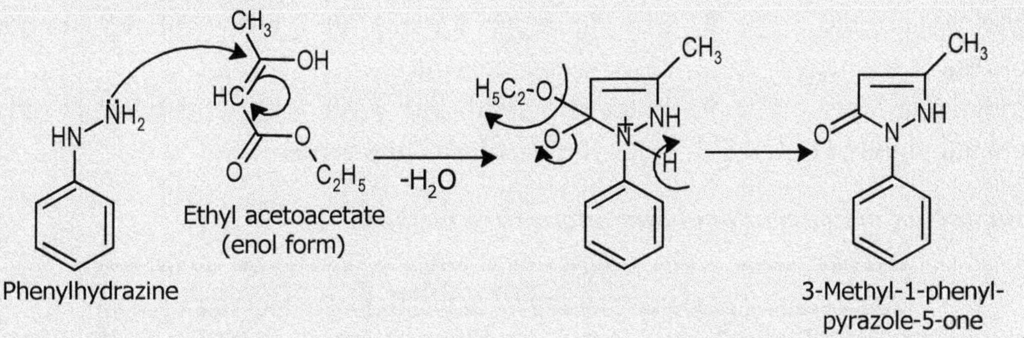

Phenylhydrazine

Ethyl acetoacetate (enol form)

$-H_2O$

3-Methyl-1-phenyl-pyrazole-5-one

Chemicals required

- Phenylhydrazine
- Ethyl acetoacetate
- Diethyl ether

Procedure:

1) Load phenylhydrazine (1 g, 9.2 mmol) and ethyl acetoacetate (1.2 g, 9.2 mmol) into microwave flask (beaker, 50 ml).
2) Place the flask/beaker into the Microwave oven and switch on the magnetic stirrer.
   - In case of Microwave synthesizer, attach the reflux condenser to the flask.
   - In case of domestic microwave oven, cover the flask with a petridish loaded with ice cubes (provides condensing effect), alternatively funnel can be used. Baker containing cold water can be used as a heating sink.
3) Irradiate the reaction mixture present in reaction flask with Microwaves at 100% intensity for 8 minutes, with intermittent cooling for 1 minute after every minute of microwave irradiation.
4) During cooling stir the reaction mixture with spatula.
5) Turn off the power supply to the Microwave oven.
6) Cool the reaction mixture, obtains after cooling.

7) Add diethyl ether (20 ml) to the reddish brown syrupy liquid.
8) Stir vigorously the mixture (yellow colored compound slowly separates out).
9) Filter the solid and wash with cold ether (5 ml) to remove the colored impurities.

Note: Monitor the progress of reaction through thin-layer chromatography (TLC).

Comparison of conventional and microwave method

	Conventional	*Microwave*
Reaction time	3 h	8 min
Reaction temperature	80-90 °C	100% intensity (850 W)
Colour	White	
Melting point	127 °C	
Percentage yield	80%	82%

Note: This reaction can be performed using acetic acid as a alternative.

Post Lab Questions:

1) What are the biproducts in this reaction?
2) What are the substitutes for phenyl hydrazine?

NUCLEOPHILIC ADDITION

Nucleophilic addition reactions are useful in the conversion of ketones into wide range of functional groups, in particular alcohol. An electron rich nucleophile (Nu) attacks electron deficient carbonyl carbon of ketones and forms single entity (addition product).

$$R-CO-R_1 + Nu \longrightarrow R-C(OH)(Nu)-R_1$$

Ketone + Nucleophile → Addition product (alcohol)

EXPERIMENT 44

Preparation of Chlorobutanol

Principle: Chlorobutanol is local anaesthetic drug. It is also used for antibacterial and preservative functions. The nucleophilic addition reaction of acetone and chloroform in the presence of potassium hydroxide produces chlorobutanol. Potassium hydroxide acts as a base catalyst.

$$H_3C-CO-CH_3 + CHCl_3 \xrightarrow[KOH]{MW\ 425\ W,\ 8\ min} H_3C-C(OH)(CH_3)-CCl_3$$

Acetone + Chloroform → Chlorobutanol

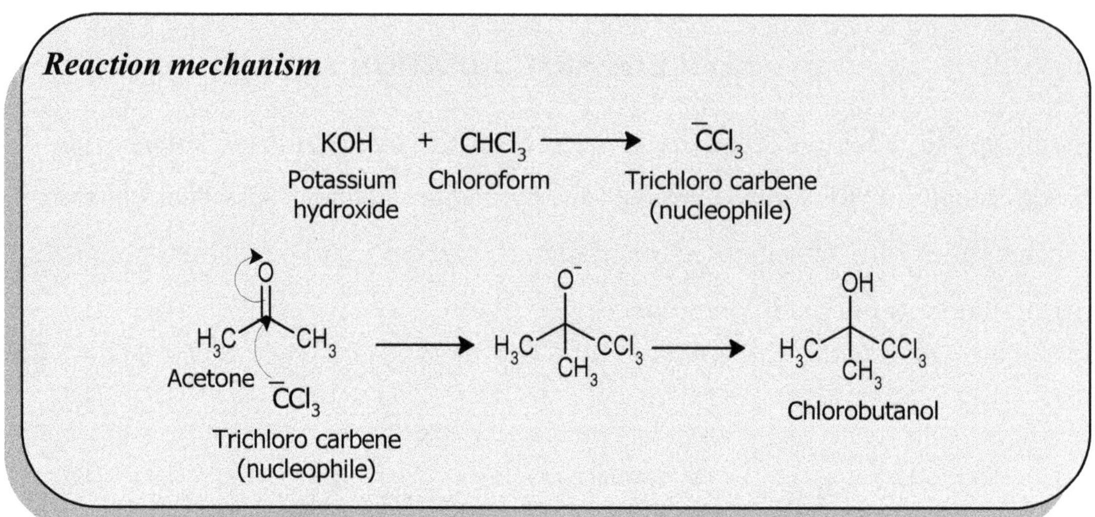

Chemicals required

- Acetone
- Chloroform
- Potassium hydroxide
- Ethanol

Procedure:

1) Load acetone (3 ml, 41 mmol), chloroform (10 ml) and potassium hydroxide (1.5 g; or sodium hydroxide) into microwave flask (beaker, 50 ml).
2) Stir the reaction mixture for 5 minutes.
3) Place the flask/beaker into the Microwave oven and switch on the magnetic stirrer.
   - In case of Microwave synthesizer, attach the reflux condenser to the flask.
   - In case of domestic microwave oven, cover the flask with a petridish loaded with ice cubes (provides condensing effect), alternatively funnel can be used. Baker containing cold water can be used as a heating sink.
4) Irradiate the reaction mixture present in reaction flask with Microwaves at 50% intensity for 8 minutes, continuously.
5) Turn off the power supply to the Microwave oven.

6) Cool the reaction mixture, filter to collects the crystals of chlorobutanol.
7) Wash the product with acetone.
8) Purify the product by recrystallization in ethanol to get pure solid chlorobutanol hemihydrate crystals.

Note: Monitor the progress of reaction through thin-layer chromatography (TLC).

Comparison of conventional and microwave method

	Conventional	***Microwave***
Reaction time	3 h	8 min
Reaction temperature	80-90 °C	50% intensity (425 W)
Colour	White	
Melting point	96-97 °C	
Percentage yield	80%	97%

Post Lab Questions:

1) What is the use of chlorobutanol?
2) What is the role of sodium hydroxide or potassium hydroxide in this reaction?

OXIDATION

In general, the term oxidation refers to the combination of oxygen with an element. Oxidation can be described on the basis of oxygen transfer, hydrogen removal and electron loss.

- In term of oxygen transfer, the process of oxygen gaining.
- In term of hydrogen transfer, the process of hydrogen loss.
- In terms of electron loss, the process of removal of electrons. This increases the oxidation state (valence) of an atom.

The oxidation process involves the breakdown of carbon-hydrogen bond and carbon forms bond with more electronegative oxygen.

EXPERIMENT 45

Preparation of Benzil

Principle: Benzil can be prepared by oxidizing the benzoin using nitric acid as oxidizing agent. Conventionally nitric acid is used in the oxidation of benzoin. Copper sulphate on alumina, ozone and zeolite produce improved yields of benzil (oxidized product of benzoin) under microwave condition. Nitric acid abstracts electron from the hydroxyl group of benzoin and generates benzil. The reaction evolves gases of oxides of nitrogen, hence precaution measure is required.

OH
O
Benzoin
HNO_3
MW 80%, 8 min
O
O
Benzil
$+ H_2O + NO_2$

Reaction mechanism

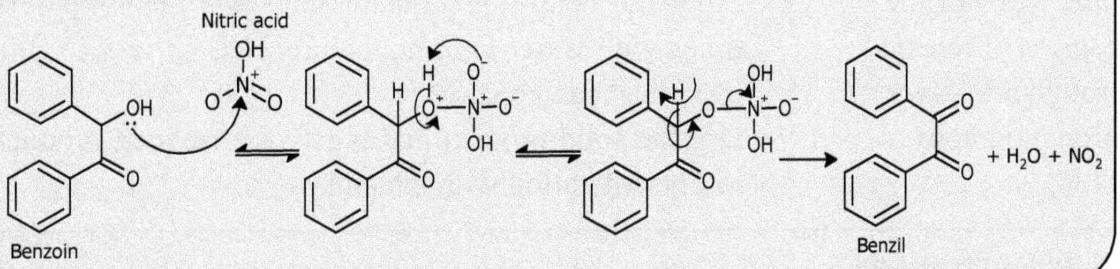

Nitric acid

Benzoin

Benzil

$+ H_2O + NO_2$

METHOD 1 (without solvent, neat reaction)

Chemicals required

- Benzoin
- Nitric acid
- Ethanol

Procedure:

1) Load benzoin (1 g, 4.71 mmol), and nitric acid (5 ml) into microwave flask (or beaker, 50 ml).
2) Place the flask into the Microwave oven and switch on the magnetic stirrer.
   - In case of Microwave synthesizer, attach the reflux condenser to the flask.
   - In case of domestic microwave oven, cover the flask with a petridish loaded with ice cubes (provides condensing effect), alternatively funnel can be used. Beaker containing cold water can be used as a heating sink.
3) Irradiate the reaction mixture present in reaction flask with Microwaves at 80% intensity for 8 minutes, with intermittent cooling for two minutes after 3 minutes of microwave irradiation.
4) During cooling stir the reaction mixture with spatula.
5) Turn off the power supply to the Microwave oven.
6) After few minutes, remove the flask from the microwave oven.

7) Cool the yellow liquid and pour into beaker containing ice-water with stirring.
8) Filter the yellow precipitate and wash the solid several times with water to remove unreacted nitric acid (litmus paper test can be used).
9) Dry the solid product and purify the isolated product by re-crystallization with ethanol.

Note: Monitor the progress of reaction through thin-layer chromatography (TLC).

Safety precaution

- The reaction produces strong nitric oxide gas (oxidation product of nitric acid).
- Protecting eyes and avoid inhalation of gas is important.

Comparison of conventional and microwave method

	Conventional	*Microwave*
Reaction time	2 h	8 min
Reaction temperature	95 °C	80% intensity (680 W)
Colour	Pale yellow	
Melting point	94-95 °C	
Percentage yield	76%	92%

METHOD 2 (using solvent)

Chemicals required

- Benzoin
- Glacial acetic acid
- Nitric acid
- Ethanol

Procedure:

1) Load benzoin (1 g, 4.7 mmol), glacial acetic acid (5 ml) and nitric acid (5 ml) into microwave flask (beaker, 50 ml).
2) Place the flask into the Microwave oven and switch on the magnetic stirrer.
    - In case of Microwave synthesizer, attach the reflux condenser to the flask.

- In case of domestic microwave oven, cover the flask with a petridish loaded with ice cubes (provides condensing effect), alternatively funnel can be used. Beaker containing cold water can be used as a heating sink.

3) Irradiate the reaction mixture present in reaction flask with Microwaves at 80% intensity for 10 minutes, with intermittent cooling for one minute after 3 minutes of microwave irradiation.
4) Turn off the power supply to the Microwave oven.
5) After few minutes, remove the flask from the Microwave oven.
6) Cool the yellow liquid and pour into beaker containing ice-water with stirring.
7) Filter the yellow precipitate and wash the solid several times with water to remove unreacted nitric acid (litmus paper test can be used).
8) Dry the solid product and purify the isolated product by re-crystallization with ethanol.

Note: Monitor the progress of reaction through thin-layer chromatography (TLC).

Alternatively this reaction can be performed using sodium nitrite as oxidizing agent in the presence of glacial acetic acid under cold conditions without microwave irradiation or heat.

Comparison of conventional and microwave method

	Conventional	***Microwave***
Reaction time	2 h	10 min
Reaction temperature	100 °C	80% intensity (680 W)
Colour	Pale yellow	
Melting point	94-95 °C	
Percentage yield	76%	92%

Post Lab Questions:

1) What types of fumes are produced in this reaction?
2) What is the role of nitric acid in the oxidation of benzoin?
3) Name the alternative oxidizing agents for this reaction.
4) Mention the purpose of petridish and beaker containing ice cold water.

EXPERIMENT 46

Preparation of 2-Nitrobenzoic Acid

Principle: Oxidation of 2-nitrotoluene by potassium permanganate produces 2-nitrobenzoic acid. Potassium hydroxide introduces alkalinity and enhances the rate of oxidation. Potassium permanganate reduces to manganese dioxide (dark precipitate), while oxidizing the 2-nitrotoluene. The dark precipitate can be dissolved by adding sodium bisulphite. This produces clear precipitate of 2-nitrobenzoic acid.

CH_3, NO_2 — 2-Nitrotoluene → (MW 850 W, 15 min; $KMnO_4$, KOH) → $COOH$, NO_2 — 2-Nitrobenzoic acid

Reaction mechanism

2-Nitrotoluene (CH_3, NO_2) → (MnO_4^-) → [Transition state: H_2C--O---MnO_3^-, NO_2] → ($- HOMnO_3^-$) → CH_2^-, NO_2 → (O) → 2-Nitrobenzoic acid ($COOH$, NO_2)

Chemicals required

- 2-Nitrotoluene
- Potassium permanganate
- Potassium hydroxide-50% (or sodium hydroxide)

Procedure:

1) Load 2-nitrotoluene (6 ml, 50 mmol), potassium permanganate (15.8 g, 100 mmol) potassium hydroxide (13 ml, 50 mmol) and water (50 ml) into the microwave flask (beaker, 50 ml).
2) Place the flask into the Microwave oven and switch on the magnetic stirrer.
   - In case of Microwave synthesizer, attach the reflux condenser to the flask.
   - In case of domestic microwave oven, cover the flask with a petridish loaded with ice cubes (provides condensing effect), alternatively funnel can be used. Beaker containing cold water can be used as a heating sink.
3) Irradiate the reaction mixture present in reaction flask with Microwaves at 680 W intensity for 15 minutes, with intermittent cooling for one minute after 5 minutes of microwave irradiation.
4) Turn off the power supply to the microwave oven.
5) After few minutes, remove the flask from the microwave oven.
6) Filter and decolorize by adding aqueous sodium bisulphite till it gets decolorized (it dissolves manganese dioxide).
7) Acidify the filtrate with concentrated hydrochloric acid or sulphuric acid (20%).
8) Filter the suspension and collect the residue, and wash with water.
9) Dry the solid residue.
10) Purify the product by re-crystallization with ethanol.

Note: Monitor the progress of reaction through thin-layer chromatography (TLC).

Comparison of conventional and microwave method

	Conventional	*Microwave*
Reaction time	3 h	15 min
Reaction temperature	100 °C	80% intensity (680 W)
Colour	Pale yellow	
Melting point	147-148 °C	
Percentage yield	76%	88%

Post Lab Questions:

1) What is the role of sodium bisulphite in this reaction?
2) Name the alternative oxidizing agents for this reaction.
3) What is the role of potassium hydroxide in this reaction?

EXPERIMENT 47

Preparation of Benzoic Acid

Principle: Oxidation of benzyl chloride by potassium permanganate produces benzoic acid. Potassium hydroxide introduces alkalinity and enhances the rate of oxidation. Potassium permanganate reduces to manganese dioxide (dark precipitate), while oxidizing the benzyl chloride. The dark precipitate can be dissolved by adding sodium bisulphite. This produces clear precipitate of benzoic acid.

Benzyl chloride $\xrightarrow[KMnO_4]{\text{MW 680 W, 15 min}}$ Benzoic acid

Reaction mechanism

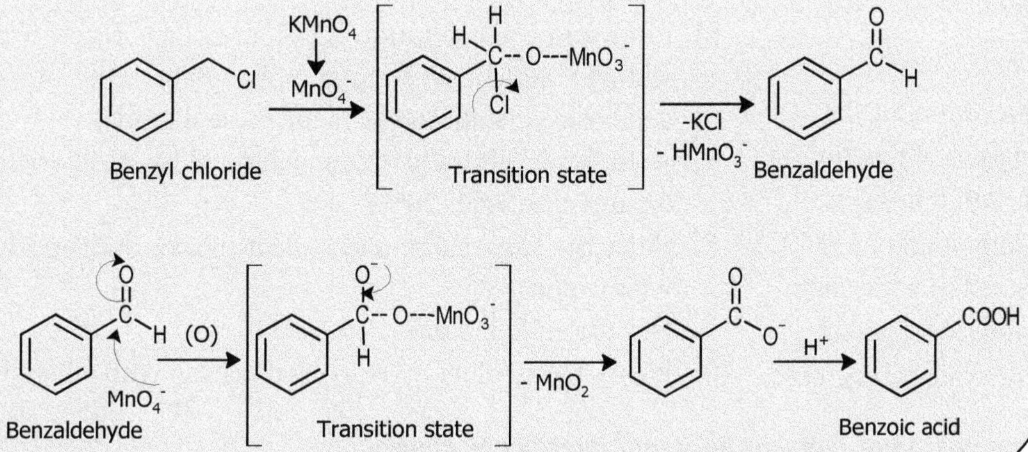

Chemicals required

- Benzyl chloride
- Potassium permanganate
- Potassium hydroxide/sodium hydroxide (50%)
- Sodium bisulphate (50%)
- Concentrated hydrochloric acid (or sulphuric acid)

Procedure:

1) Load benzyl chloride (6 ml, 52.14 mmol), potassium permanganate (15.8 g, 100 mmol) potassium hydroxide (13 ml) and water (50 ml) into the reaction flask.
2) Place the flask into the Microwave oven and switch on the magnetic stirrer.
   - In case of Microwave synthesizer, attach the reflux condenser to the flask.
   - In case of domestic microwave oven, cover the flask with a petridish loaded with ice cubes (provides condensing effect), alternatively funnel can be used. Beaker containing cold water can be used as a heating sink.
3) Irradiate the reaction mixture present in reaction flask with Microwaves at 680 W intensity for 15 minutes, with intermittent cooling for one minute after 5 minutes of microwave irradiation.

4) Turn off the power supply to the microwave oven.
5) After few minutes, remove the flask from the microwave oven.
6) Filter and if the solution is coloured then decolorize by adding aqueous sodium bisulphite till it gets decolorized (it dissolves manganese dioxide).
7) Acidify the filtrate with concentrated hydrochloric acid or sulphuric acid (20%).
8) Filter the suspension and collect the residue, and wash with water.
9) Dry the solid residue.
10) Purify the product by re-crystallization with ethanol.

Note: Monitor the progress of reaction through thin-layer chromatography (TLC). Instead of potassium hydroxide, sodium hydroxide can be used.

Comparison of conventional and microwave method

	Conventional	*Microwave*
Reaction time	3 h	15 min
Reaction temperature	100 °C	80% intensity (680 W)
Colour	White	
Melting point	112 °C	
Percentage yield	76%	88%

Post Lab Questions:

1) What is the role of sodium bisulphite in this reaction?
2) Name the alternative oxidizing agents for this reaction.
3) What is the role of potassium hydroxide in this reaction?

HYDROLYSIS

Splitting of an ester or amide into corresponding acids is an typical example of hydrolysis. Acid or base catalysed hydrolyses are common.

- Ester hydrolysis generates acid and alcohol.

$$\underset{\text{Ester}}{R\text{-}COO\text{-}R^1} + H_2O \longrightarrow \underset{\text{Acid}}{R\text{-}COOH} + \underset{\text{Alcohol}}{R^1\text{-}OH}$$

- Amide hydrolysis generates acid and amine (or ammonia).

$$\underset{\text{Amide}}{R\text{-}CO\text{-}NH_2} + H_2O \longrightarrow \underset{\text{Acid}}{R\text{-}COOH} + \underset{\text{Ammonia}}{NH_3}$$

Water autoionizes into negative hydroxyl ion and hydrogen ion. In general, the hydrolysis reaction is mediated by base (sodium hydroxide or potassium hydroxide). Base (hydroxyl ion) converts the acid component of ester or amide into corresponding salt, which favours the forward reaction (irreversible). Hydrolysis occurs, when the nucleophile attacks the carbon of the carbonyl group (ester or amide). In an aqueous base, hydroxyl ions are better nucleophile. In acid, the carbonyl group becomes protonated and this leads to a much easier nucleophilic attack. Hydrolysis of triglycerides (esters of long chain fatty acids) by base is known as saponification.

EXPERIMENT 48

Preparation of Benzoic Acid

Principle: Primary amides upon reaction (hydrolysis) with warm dilute sodium hydroxide (or dilute sulphuric acid) produce the sodium salt of corresponding acid and releases ammonia. The reaction (hydrolysis) can be driven to completion by boiling the reaction mixture. Benzamide (primary amide) undergoes hydrolysis with warm sodium hydroxide (10%) solution and produces sodium benzoate (sodium salt of benzoic acid). The reaction also liberates ammonia gas. The sodium salt of benzoic acid is soluble, hence acidification with hydrochloric acid is required to precipitate the benzoic acid.

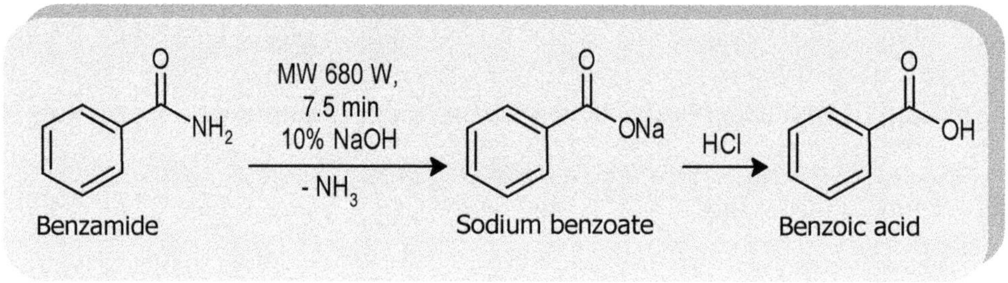

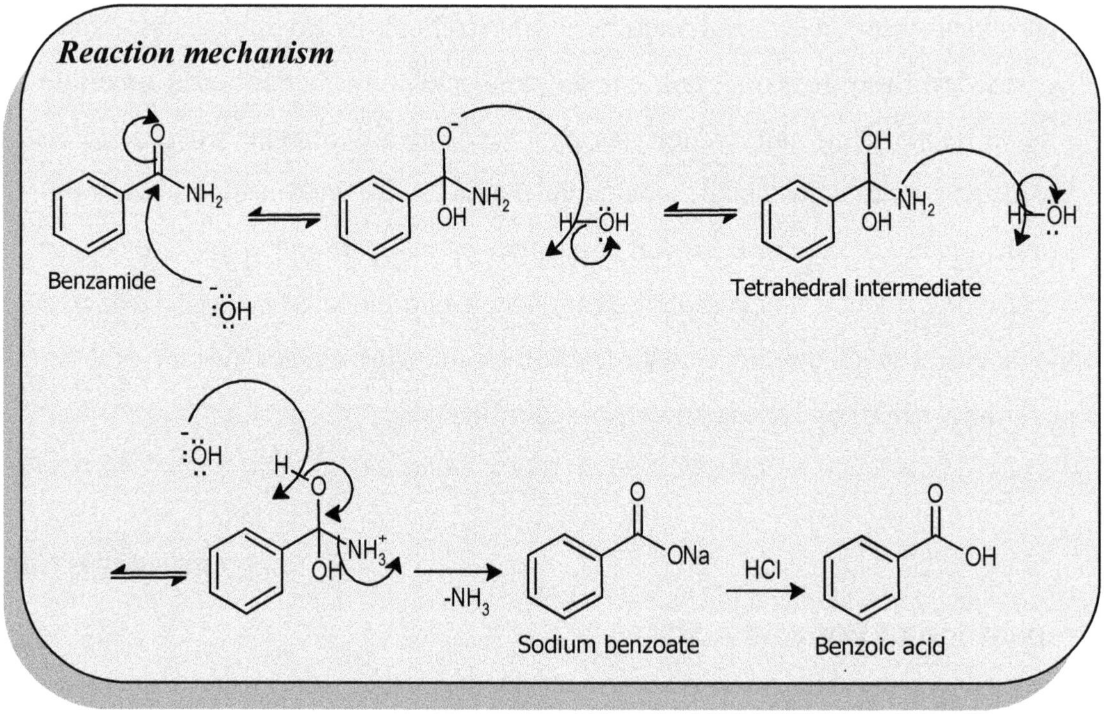

Chemicals required

- Benzamide
- Dilute sodium hydroxide (10%)
- Hydrochloric acid

Procedure:

1) Load benzamide (1.0 g, 8.25 mmol) and sodium hydroxide (10%, 15 ml) into microwave flask (beaker, 50 ml).
2) Place the flask into the Microwave oven and switch on the magnetic stirrer.
   - In case of Microwave synthesizer, attach the reflux condenser to the flask.
   - In case of domestic microwave oven, cover the flask with a petridish loaded with ice cubes (provides condensing effect), alternatively funnel can be used. Baker containing cold water can be used as a heating sink.
3) Irradiate the reaction mixture at 80% intensity for 7.5 minutes, with intermittent cooling for one minute after 3 minutes of microwave irradiation.
4) Turn off the power supply to the microwave oven.
5) After few minutes, remove the flask from the microwave oven.
6) Cool the reaction mixture on ice water bath.
7) Acidify with concentrated hydrochloric acid to precipitate the benzoic acid.
8) Filter the precipitated benzoic acid and wash it several times with cold water to remove unreacted acid.
9) Dry the isolated benzoic acid and purify the product by recrystallization with hot water.

Note: Monitor the progress of reaction through thin-layer chromatography (TLC).

Comparison of conventional and microwave method

	Conventional	*Microwave*
Reaction time	30 min	7.5 min
Reaction temperature	100 °C	80% intensity (680 W)

Table *Contd...*

Colour	White	
Melting point	122-123 °C	
Percentage yield	98%	98%

Post Lab Questions:

1) Why the product is washed with cold water?
2) What are the different types of hydrolysis reactions performed in laboratories?
3) Mention the different hydrolysis reactions.
4) What is the role of hydrochloric acid in this hydrolytic reaction.

EXPERIMENT 49

Preparation of Benzoic Acid from Phenyl Benzoate

Principle: Esters upon base hydrolysis generates corresponding acid in salt form. The hydrolysis can be promoted by using acids and bases. Phenyl benzoate (ester) upon alkaline hydrolysis yields sodium benzoate (sodium salt of acid). The sodium salt (insoluble) upon acidification precipitates benzoic acid (insoluble).

Phenyl benzoate → (MW 850 W, 4 min, 10% NaOH) → Sodium benzoate (ONa) → (H^+) → Benzoic acid (OH)

Reaction mechanism

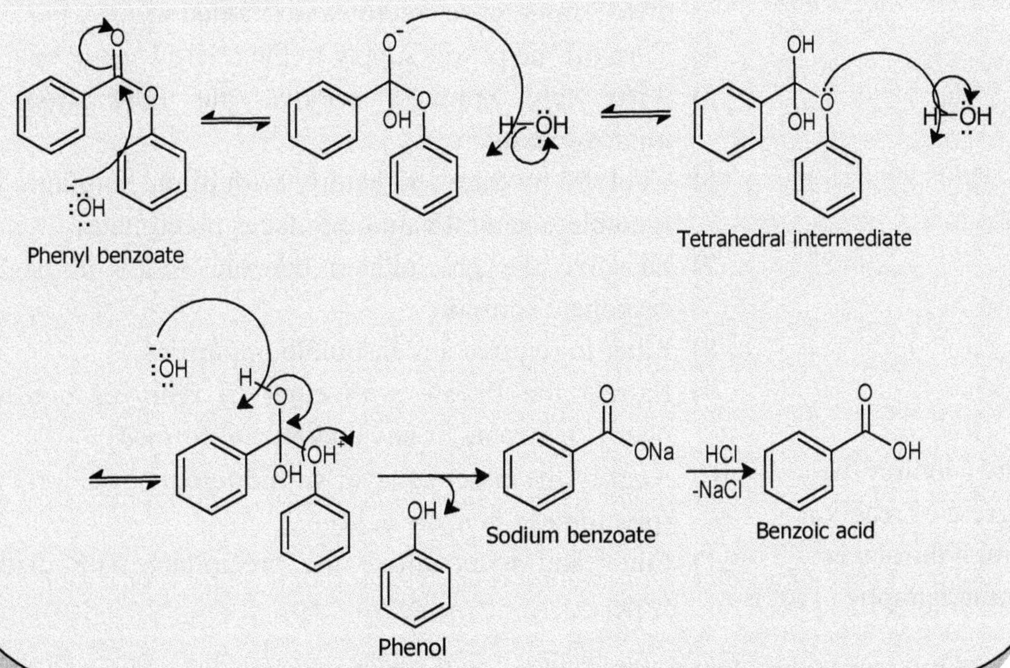

Phenyl benzoate

Tetrahedral intermediate

Phenol

Sodium benzoate

HCl

-NaCl

Benzoic acid

Chemicals required

- Phenyl benzoate
- Sodium hydroxide (10%)
- Dilute sulphuric acid
- Sodium carbonate solution (10%)
- Dilute hydrochloric acid
- Ether
- Methanol

Procedure:

1) Load phenyl benzoate (0.5 g, 2.5 mmol), sodium hydroxide solution (10%, 7.5 ml) and methanol (6 ml) into microwave flask (beaker, 50 ml).
2) Place the flask into the Microwave oven and switch on the magnetic stirrer.
   - In case of Microwave synthesizer, attach the reflux condenser to the flask.
   - In case of domestic microwave oven, cover the flask with a petridish loaded with ice cubes (provides condensing effect), alternatively funnel can be used. Baker containing cold water can be used as a heating sink.

3) Irradiate the reaction mixture at 100% intensity (850 W) for 4 minutes, with intermittent cooling for one minute after 3 minutes of microwave irradiation.
4) Turn off the power supply to the microwave oven.
5) After few minutes, remove the flask from the microwave oven.
6) Cool the mixture and acidify with dilute sulphuric acid (complete neutralization produces precipitate).
7) Dissolve the precipitated benzoic acid with sodium carbonate solution
8) Filter to remove any insoluble impurities.
9) Extract the filtrate with ether (it removes unreacted phenyl benzoate, if any and phenol formed).
10) Acidify the aqueous layer with dilute hydrochloric acid (precipitates benzoic acid).
11) Filter and recrystallize the precipitate with distilled water.

Note: Monitor the progress of reaction through thin-layer chromatography (TLC).

Comparison of conventional and microwave method

	Conventional	*Microwave*
Reaction time	30 min	4 min
Reaction temperature	50-60 °C	100% intensity (850 W)
Colour	White	
Melting point	122-123 °C	
Percentage yield	98%	98%

Post Lab Questions:

1) What is the role of methanol and solvent ether in this preparation?
2) Name the precipitate formed after acidification with sulphuric acid?
3) What is the role/importance of sodium carbonate in this reaction?

PART - 3

QUALITATIVE ANALYSIS

EXPERIMENT 50

Qualitative Analysis of Carbohydrates

Principle: Carbohydrates (hydrates of carbons) are composed of carbon, hydrogen and oxygen ($C_m(H_2O)_n$). Carbohydrates are essential for the energy metabolism and provide energy (ATP) to the living cells. Chemically these are defined as polyhydroxy alcohols or polyhydroxy ketones or precursors of these. Carbohydrates can be grouped into reducing (eg: maltose) and non-reducing sugars (eg: sucrose). Reducing sugars can be identified through Fehling's, Benedict,s, Barfoed's, Tollen's and Osazone tests.

Chemicals required

- Sample
- Fehling's reagent
- Benedict's reagent

Procedure:

1) Prepare aqueous sample solutions of fructose, glucose, maltose and lactose.
2) Transfer specified quantity of the sample into beakers (25 ml).
3) Place the flask into the Microwave oven and switch on the magnetic stirrer.
4) Irradiate the mixture (specified amount) with microwaves at 80% intensity for appropriate duration.
5) Turn off the power supply to the Microwave oven.
6) Observe the formation of colored solution/precipitate.

Comparison of conventional and microwave method

	Qualitative test	*Observation*	*Conventional*	*Microwave, 80% (seconds)*
1	**Benedicts Test:** Sample solution (5 ml) + Benedicts reagent (1 ml)			
	• Fructose	Red precipitate	60	40
	• Glucose		105	40
	• Maltose		150	30
	• Lactose		120	50
2	**Fehling test:** Sample solution (2 ml) + Fehling's A (1 ml) and B (1 ml) reagent			
	• Fructose	Red precipitate	50	40
	• Glucose		70	40
	• Maltose		90	50
	• Lactose		100	50
3	**Tollen's Test**: Sample solution (1 ml) + Tollen's reagent (1 ml)			
	• Fructose	Silver mirror	60	25
	• Glucose		70	30
	• Maltose		75	40
	• Lactose		90	40
4	**Barfoed's Test**: Sample solution (2 ml) + Barfoed's reagent (2 ml)			
	• Fructose	Red precipitate	90	25
	• Glucose		120	30
	• Maltose		75	40
	• Lactose		90	40
5	**Osazone's test**: Sample solution (1 gm in 5 ml) + glacial acetic acid (2 ml) + phenylhydrazine (2 ml)			
	• Glucose	Needle-shaped crystals	120	60
	• Fructose	Needle-shaped crystals	120	60
	• Lactose	Cotton-ball shaped crystals	480	75
	• Maltose	Sun flower shaped crystals	480	75

Post Lab Questions:

1) What is the composition of following reagents?
   - Benedict's reagent.
   - Fehling solution A and B.
2) What are reducing sugars?
3) Why silver mirror is observed in Tollen's test?

Preparation of Derivatives of Organic Compounds

Principle: The chemical structure elucidation is very important in natural product and synthetic chemistry. The molecular frame work can be established utilizing the nature of functional groups. The chemical behavior and the functional groups associated with the molecules can be revealed through molecular derivatization. The molecular characterization and identification can be achieved through suitable solid derivatization and further analysis. Accurate determination physical constants such melting point and boiling point of the derivative products is useful in structural determination. The chemical properties of the molecule under study decide the type of derivative.

1) Nitration Test:

$$\text{Toluene} \xrightarrow[H_2SO_4,\ HNO_3]{\text{MW 850 W}} \text{4-Nitrotoluene}$$

Chemicals required

- Sample
- Sulfuric acid
- Nitric acid
- Ethanol

Procedure:

1) Transfer the sample into the reaction flask.
2) Add sulfuric acid and nitric acid (required quantities) and mix it well.
3) Place the flask into the Microwave oven and switch on the magnetic stirrer.
    - In case of domestic microwave oven, cover the flask with a petridish loaded with ice cubes

(provides condensing effect), alternatively funnel can be used. Beaker containing cold water can be used as a heating sink.

4) Irradiate the mixture (specified amount) with microwaves at 100% intensity for appropriate duration with intermittent cooling for 2 min after every minute of microwave irradiation.
5) Turn off the power supply to the Microwave oven.
6) Pour the reaction mixture into water (20 ml) contained in beaker with stirring.
7) Filter the precipitate (nitrated derivative) formed and recrystalize with ethanol.
8) Determine the melting point formed the derivative formed.

Comparison of conventional and microwave method

Sample (0.5 ml)	***Nitration mixture (ml)***	***Product precipitated (colour)***	***Conventional***		***Microwave, 100%***	
			Duration (seconds)	***m.p (°C)***	***Duration (sec)***	***m.p (°C)***
Toluene	5	4-Nitrotoluene (Pale yellow)	120	53	20	53-54
Naphthalene	5	Nitro-naphthalene (Brown)	300	59	60	59-60
Chlorobenzene	4	Nitro-chlorobenzene (Pale yellow)	600	52	80	52

2) Oxime Test:

Benzaldehyde —(MW 850 W; $NH_2OH.HCl$, Pyridine)→ Benzaldoxime + HCl + H_2O

Chemicals required

- Sample
- Hydroxylamine hydrochloride
- Methanol
- Pyridine

Procedure:

1) Transfer the sample (0.5 g), hydroxylamine hydrochloride (0.5 g), methanol (8 ml) and pyridine (0.2 ml) into the reaction flask and mix it well.
2) Place the flask into the Microwave oven and switch on the magnetic stirrer.
   - In case of domestic microwave oven, cover the flask with a petridish loaded with ice cubes (provides condensing effect), alternatively funnel can be used. Beaker containing cold water can be used as a heating sink.
3) Irradiate the mixture (specified amount) with microwaves at 100% intensity for appropriate duration, with intermittent cooling for 2 min after every minute of microwave irradiation.
4) Turn off the power supply to the Microwave oven.
5) Pour the reaction mixture into water (20 ml) contained in beaker with stirring.
6) Filter the precipitate (nitrated derivative) formed and recrystallize with ethanol.
7) Determine the melting point formed the derivative formed.

Comparison of conventional and microwave method

Samples (0.5 g)	*Product precipitated (colour)*	*Conventional*		*Microwave, 100%*	
		Duration (minutes)	*m.p (°C)*	*Duration (minutes)*	*m.p (°C)*
Acetaldehyde	Pale yellow	30	46	2	47
Benzaldehyde	Pale white	30	35	2.5	35
Acetone	Pale white	30	59	1	59
Benzophenone	Pale white	30	142	1.2	143

3) Hydrazone Test:

Chemicals required

- Sample
- Phenyl hydrazine hydrochloride
- Sodium acetate
- Water

Procedure:

1) Transfer the sample (0.4 g), phenyl hydrazine hydrochloride (0.5 g), sodium acetate (0.8 g) and water (5 ml) into the reaction flask and mix it well.
2) Place the flask into the Microwave oven and switch on the magnetic stirrer.
   - In case of domestic microwave oven, cover the flask with a petridish loaded with ice cubes (provides condensing effect), alternatively funnel can be used. Beaker containing cold water can be used as a heating sink.
3) Irradiate the mixture (specified amount) with microwaves at 100% intensity for appropriate duration, with intermittent cooling for 2 min after every minute of microwave irradiation.
4) Turn off the power supply to the Microwave oven.
5) Pour the reaction mixture into water (20 ml) contained in beaker with stirring.
6) Filter the precipitate (nitrated derivative) formed and recrystalize with ethanol.
7) Determine the melting point formed the derivative formed.

Comparison of conventional and microwave method

Sample (0.4 g)	*Product precipitated (colour)*	*Conventional*		*Microwave, 100%*	
		Duration (seconds)	*m.p (°C)*	*Duration (seconds)*	*m.p (°C)*
Benzaldehyde	White	900	157	60	158
Acetone	White	900	42	40	42
Benzophenone	White	900	135	40	135

Post Lab Questions:

1) What is derivatization?
2) How derivatization is useful in the qualitative analysis?

Degradation of Atropine

Principle: Atropine is an alkaloid obtained from the roots and leaves of *Atropa belladonna* (Family: Solanaceae). Chemically it is a tropyl ester of tropic acid. Atropine is a racemic product formed during the extraction of hyoscyamine. The alkaline hydrolysis of atropine releases tropine and tropic acid. Conventional methods require heating for 45 minutes. Microwave method requires 1 minute 40 seconds of microwave irradiation.

Atropine $\xrightarrow[Ba(OH)_2]{\text{MW 850 W, 2 min}}$ Tropine + Tropic acid

Chemicals required

- Atropine
- Barium hydroxide (10%)
- Chloroform

Procedure:

1) Transfer atropine (0.2 g) into reaction flask (100 ml) and add barium hydroxide (7.5 ml; 10%).
2) Place the flask into the Microwave oven and switch on the magnetic stirrer.
   - In case of Microwave synthesizer, attach the reflux condenser to the flask.
   - In case of domestic microwave oven, cover the flask with a petridish loaded with ice cubes (provides condensing effect), alternatively funnel can be used. Beaker containing cold water can be used as a heating sink.

3) Irradiate the reaction mixture present in reaction flask with Microwaves at 100% intensity for 2 minutes.
4) Turn off the power supply to the Microwave oven.
5) After few minutes (allow it to cool), remove the flask from the Microwave oven.
6) Extract the reaction mixture with chloroform (5 ml) three times.
7) Tropine and tropic acid (degradation product) enters into chloroform layer (non-aqueous layer) from aqueous layer.
8) Combine the chloroform extracts and evaporate the solvent to dryness.
9) The degradation products can be detected through chemical identification tests.

Comparison of conventional and microwave method

	Conventional	*Microwave*
Reaction time	45 min	2 min
Reaction temperature	95 °C	100% intensity (850 W)

Conformation Tests for Degradation

The chemical tests are used to confirm the presence of an alkaloid, atropine as well as their degradation products.

- **Test for Tropine:** To the sample add chromium trioxide (1 g) concentrated sulphuric acid (1 ml) and water (25 ml).
  - Greenish blue precipitate indicates the presence of tropine.
- **Test for Tropic acid:** To the sample add dilute ammonia solution. Boil (removes unreacted ammonia), dilute with water and add ferric chloride solution.
  - A buff colour precipitate indicates the presence of tropic acid.

Post Lab Questions:

1) Name the botanical source of atropine.
2) How will you identify the degradation products of atropine through chemical analysis?
3) What is the role of chloroform in this experiment?

EXPERIMENT 53

Degradation of Trimyristin

Principle: Trimyristin is an tiglycerides obtained from the roots and leaves of *Myristica fragrans* (Family: Myristicaceae). It is glyceryl ester of Myristic acid. Trimyristin on alkaline hydrolysis (excessive alkaline) liberates myristic acid and glycerol. It forms alkaline salt of myristic acid. The salt upon acidification with hydrochloric acid precipitates myristic acid. It occurs as white to yellowish grey solid with a melting point of 56-57 °C. The microwave product is very pure, when compared with the conventional product.

$$\begin{array}{l} H_2C-O-CO-C_{13}H_{17} \\ \;\;| \\ HC-O-CO-C_{13}H_{17} \\ \;\;| \\ H_2C-O-CO-C_{13}H_{17} \end{array} \xrightarrow[KOH,\ C_2H_5OH]{MW\ 850\ W,\ 5\ min} \begin{array}{l} H_2C-OH \\ \;\;| \\ HC-OH \\ \;\;| \\ H_2C-OH \end{array} + 3\ C_{13}H_{17}COOK$$

Trimyristin → Glycerol + Potassium salt of Myristic acid

$$C_{13}H_{170}COOK + HCl \longrightarrow C_{13}H_{170}COOH + KCl$$

Potassium salt of Myristic acid → Myristic acid

Chemicals required

- Trimyristin
- Potassium hydroxide (ethanolic; 3.5%)
- hydrochloric acid

Procedure:

1) Transfer trimyristin (2 g, 3 mmol) into reaction flask (100 ml) and ethanolic potassium hydroxide (30 ml, 3.5%) into a reaction flask.
2) Place the flask into the Microwave oven and switch on the magnetic stirrer.
   - In case of Microwave synthesizer, attach the reflux condenser to the flask

- In case of domestic microwave oven, cover the flask with a petridish loaded with ice cubes (provides condensing effect), alternatively funnel can be used. Baker containing cold water can be used as a heating sink. It absorbs excessive energy.

3) Irradiate the reaction mixture present in reaction flask with Microwaves at 100% intensity for 5 minutes, with intermittent cooling for 2 min after every minute of microwave irradiation.
4) Turn off the power supply to the Microwave oven.
5) After few minutes (allow it to cool), remove the flask from the Microwave oven.
6) Filter the reaction mixture to remove the insoluble impurities, if any present.
7) Acidify the filtrate with dilute hydrochloric acid under cold condition to precipitate the myristic acid.
8) Filter and wash the precipitate with cold water.
9) Recrystallize the product with ethanol.

Comparison of conventional and microwave method

	Conventional	*Microwave*
Reaction time	30 min	5 minutes
Reaction temperature	45 °C	100% intensity (850 W)
Melting point	52 - 53 °C	53 °C
Yield	98%	100%

Post Lab Questions:

1) What are the degradation products of trimyristin? Give their structures.

PART - 4

QUANTITATIVE ANALYSIS

Determination of Saponification Value

Principle: The number of milligram of potassium hydroxide, which neutralizes the free fatty acid obtained from the hydrolysis of fat or oil (1 g) is known as saponification value. This value can be used as a measure of the quality of fats and oils. It is also important in the determination of HLB values of surfactants. Each fat/oil molecule requires three molecules of potassium hydroxide for neutralization of liberated acid (produces potassium salt). Hence, the molecular weight of the sample can be determined from saponification value. The excess of alkali can be determined by back titration (residual titration) with standard acid. The volume of alkali consumed for the hydrolysis can be calculated from the blank titration. This volume is useful in determining the saponification value. Conventional methods require heating for 2-3 hours. Microwave method requires 15 minutes of irradiation. The saponification values of oils and surfactants can be determined using microwave irradiations (alternative for conventional reflux procedure). This microwave procedure is rapid, simple and accurate.

$$\underset{\text{Oil/ fat}}{R{-}\overset{O}{\overset{\|}{C}}{-}O{-}R1} \xrightarrow[\text{KOH, } C_2H_5OH]{\text{MW 510 W, 15 min}} \underset{\text{Soap}}{R{-}\overset{O}{\overset{\|}{C}}{-}O{-}K} + \underset{\text{Alcohol}}{R_1{-}OH}$$

Chemicals required

- Sample (oil/surfactant)
- Potassium hydroxide (ethanolic, 0.5 N)
- Hydrochloric acid (0.5 N)

Procedure:

1) Transfer the sample (oil or surfactant; 2 g) into microwave flask (conical flask, 100 ml) and add ethanolic potassium hydroxide (25 ml, 0.5 N).
2) Stir the mixture well, place a few boiling chips (to avoid bumping while heating).

- Phenolphthalein

3) Place the flask into the Microwave oven and switch on the magnetic stirrer.
   - In case of Microwave synthesizer, attach the reflux condenser to the flask.
   - In case of domestic microwave oven, cover the flask with a petridish loaded with ice cubes (provides condensing effect), alternatively funnel can be used. Beaker containing cold water can be used as a heating sink.
4) Irradiate the reaction mixture (excessive unreacted potassium hydroxide) present in reaction flask with Microwaves at 60% intensity for 15 minutes, with intermittent cooling for 1 minute after every 2 minutes of microwave irradiation.
5) During cooling stir the reaction mixture with spatula.
6) Turn off the power supply to the Microwave oven.
7) After few minutes (allow it to cool), remove the flask from the Microwave oven.
8) Titrate the reaction mixture against hydrochloric acid (0.5 N) using phenolphthalein as an indicator.
9) Appearance of pale pink colour indicates the endpoint and note down the volume run down (A).
10) Perform blank titration using potassium hydroxide (25 ml, 0.5 N) omitting the sample.
11) The saponification value of the sample can be calculated using the formula given below.

$$\text{Saponification value} = \frac{28.05\,(B - A)}{W} \times 100$$

B = Volume in blank

A = Volume in assay

W = Weight of the sample

Comparison of conventional and microwave method

	Conventional	*Microwave*
Reaction time	30 min	15 min
Reaction temperature	95 °C	60% intensity

Saponification values of oils and surfactants by Conventional and Microwave methods

Sample	*Standard saponification values*	*Saponification values*	
		Conventional (30 minutes)	*Microwave, 60% (5 minutes)*
Oils			
Castor oil	176 - 187	179.63	182.39
Arachis oil	185 - 196	189.00	188.79
Coconut oil	255 - 258	256.55	255.37
Gingili oil	188 - 193	190.59	188.18
Sunflower oil	188 - 194	191.84	190.62
Neem oil	196 - 200	197.20	198.30
Mustard oil	170 - 176	171.90	172.80
Soya bean oil	189 - 195	189.50	189.30
Surfactants			
Tween 20	40 - 55	49.50	50.06
Tween 40	41 - 52	50.38	50.44
Tween 60	45 - 55	48.20	52.92
Tween 80	45 - 55	49.00	52.09
Span 20	158 - 170	163.93	162.50
Span 40	140 - 150	146.70	143.58
Span 60	147 - 157	150.57	152.53
Span 65	176 - 188	181.02	179.61

Post Lab Questions:

1) What is the significance of saponification value in the oil and fat analysis?
2) What are the degradation products of oil/fat after alkaline hydrolysis?
3) Explain the principle involved in the determination of saponifcation value.
4) Why blank titration is required in this determination?

EXPERIMENT 55

Determination of Loss On Drying

Principle: The loss of weight expressed in percentage (%) W/W (%w/w), resulting from loss of water and volatile matter is known as loss on drying (LOD). It does not refer the molecularly bound water (water for crystallization). This test is useful in measuring the amount of water and volatile matter in a sample under specific conditions. Loss on drying is an important quality control parameter in the Pharmacopoeias. It can be determined by heating the sample below its melting point in an oven. In general, the LOD values are given in ranges. The degree of drying depends on temperature and drying time. The microwave method requires heating for fewer minutes, whereas the conventional method requires heating for hours (~3 hour).

Procedure

1) Transfer the sample (weight specified in the pharmacopoeia) into the weighing bottle/crucible.
   - Note down the weight of the empty weighing bottle/crucible before transferring the sample (W_1).
   - Reduce the sample size not more than 2 mm diameter by quick crushing before weighing, if necessary.
   - Note down the weight of the weighing bottle/crucible along with the sample (W_2).
2) Place the weighing bottle/crucible into the Microwave oven and switch on the magnetic stirrer.

3) Irradiate the sample present in weighing bottle/crucible with Microwaves at 100% intensity for appropriate time, with intermittent cooling for 2 min after every minute of microwave irradiation.

   Stopper is not required while irradiating with microwaves. It can be placed after removing from the oven.
4) Turn off the power supply to the Microwave oven.
5) After few minutes (allow it to cool), remove the weighing bottle/crucible from the Microwave oven.
6) Weigh the weighing bottle/crucible after irradiation (W_3).
7) The percentage loss on drying for the sample can be calculated using the formula given below.

$$\%\ LOD = \frac{W_2 - W_3}{W_2 - W_1} \times 100$$

W_1 = Weight of the empty bottle/crucible

W_2 = Weight of the empty bottle/crucible with sample before drying

W_3 = Weight of the empty bottle/crucible with sample after drying

Comparison of conventional and microwave method

Sample	***% LOD***	
	Conventional method (3 hours at 10 ºC)	***Microwave method, 100% (minutes)***
Potassium iodide	0.89	0.89 (9)
Sodium salicylate	0.35	0.35 (25)
Sodium benzoate	4.04	4.02 (5)

Post Lab Questions:

1) What is loss on drying (LOD)?
2) Mention the pharmaceutical significance of LOD?

Estimation of Amides

Principle: Alkaline hydrolysis of amides liberates amines and acids with an aid of heating. The alkaline hydrolysis process utilizes excess of alkali. The excess of alkali can be determined by back titration (residual titration) with standard acid. The volume of alkali consumed for the hydrolysis can be calculated from the blank titration. This volume is useful in calculating the amount of amides. Conventional methods require heating for 2-3 hours. Microwave method requires 20 minutes of microwave irradiation. This microwave method conserves energy and reaction duration. Hence this method is eco-friendly.

$$\underset{\text{Amide}}{R{-}\overset{\overset{\displaystyle O}{||}}{C}{-}NH_2} \xrightarrow[\text{NaOH, } CH_3OH]{\text{MW 510 W}} \underset{\text{Sodium salt of carboxylate}}{R{-}\overset{\overset{\displaystyle O}{||}}{C}{-}ONa} + \underset{\text{Ammonia}}{NH_3}$$

$$\text{NaOH (residual)} + HCl \longrightarrow NaCl + H_2O$$

Chemicals required

- Sample
- Methanol
- Sodium hydroxide (ethanolic; 2 N)
- Hydrochloric acid (1 N)
- Phenolphthalein indicator

Procedure:

1) Transfer the amide sample (0.5 g) into microwave flask (conical flask, 100 ml).
2) Dissolve the sample with methanol (5 ml) and add standard sodium hydroxide (20 ml; 2 N).
3) Place the flask into the Microwave oven and switch on the magnetic stirrer.
    - In case of Microwave synthesizer, attach the reflux condenser to the flask.

- In case of domestic microwave oven, cover the flask with a petridish loaded with ice cubes (provides condensing effect), alternatively funnel can be used. Beaker containing cold water can be used as a heating sink.

4) Irradiate the reaction mixture present in reaction flask with Microwaves at 60% intensity for appropriate duration to effect amide hydrolysis, with intermittent cooling for 1 minute after every 5 minutes of microwave irradiation. During cooling stir the reaction mixture with spatula.
5) Turn off the power supply to the Microwave oven.
6) After few minutes (allow it to cool), remove the flask from the Microwave oven.
7) Titrate the excessive (unreacted) sodium hydroxide (back titration) with hydrochloric acid (1 N) using phenolphthalein as an indicator.
8) Perform blank titration using sodium hydroxide (20 ml; 2 N), omitting the sample.
9) The percentage purity of amide in the sample can be calculated using the formula given below.

   % Purity =

$$\frac{\text{M Wt of sample} \times \text{Blank} - \text{Assay} \times \text{Normality of HCl} \times 100}{1000 \times \text{Weight of the sample}}$$

Comparison of conventional and microwave method

Sample	***% Purity***	
	Conventional method (3 hours)	***Microwave method, 60% (minutes)***
Acetamide	79.89	91.14 (20)
Benzamide	99.83	99.92 (20)
Formamide	99.43	99.67 (15)

Post Lab Questions:

1) Explain the term residual titration.
2) What are the advantages of microwave methods?

Estimation of Esters

Principle: The organic esters can be estimated by hydrolysis with known and excess volume of standard alkali. The alkali hydrolysis reaction generates corresponding acid and alcohol. This alkaline hydrolysis of ester is known as saponification. The excess of alkali can be determined by back titration (residual titration) with hydrochloric acid. The volume of alkali consumed for the hydrolysis can be calculated from the blank titration. This volume is useful in calculating the amount of esters. Conventional methods require heating for ~3 hours. Microwave method requires less than 10 minutes of microwave irradiation.

Saponification equivalent of an ester is defined as the weight of ester (in g), which reacts with 1 g of equivalent of a strong base.

$$RCOOR_1 \xrightarrow[KOH,\ CH_3OH]{MW\ 850\ W} RCOOK + R_1{-}OH$$

Ester → Potassium salt of carboxylate + Alcohol

$$KOH\ (residual) + HCl \longrightarrow KCl + H_2O$$

Chemicals required

- Sample
- Methanol
- Potassium hydroxide solution (ethanolic; 0.5 N)
- Hydrochloric acid (0.5 N)

Procedure:

1) Transfer the sample (0.5 g) into microwave flask (conical flask, 100 ml)..
2) Dissolve the sample with methanol (5 ml) and add potassium hydroxide (ethanolic; 50 ml, 0.5 N).
3) Place the flask into the Microwave oven and switch on the magnetic stirrer.

- Phenolphthalein indicator

  - In case of Microwave synthesizer, attach the reflux condenser to the flask.
  - In case of domestic microwave oven, cover the flask with a petridish loaded with ice cubes (provides condensing effect), alternatively funnel can be used. Beaker containing cold water can be used as a heating sink.

4) Irradiate the reaction mixture present in reaction flask with Microwaves at 100% intensity for appropriate duration to effect ester hydrolysis, with intermittent cooling for 1 minute after every minute of microwave irradiation.
5) Turn off the power supply to the Microwave oven.
6) After few minutes (allow it to cool), remove the flask from the Microwave oven.
7) Titrate the excess (unreacted) of potassium hydroxide (back titration) with hydrochloric acid (0.5 N) using phenolphthalein as an indicator.
8) Perform blank titration using methanol and ethanolic potassium hydroxide (50 ml; 0.5 N) omitting the sample.
9) The percentage purity of ester in the sample can be calculated using the formula given below.

   % Purity =

$$\frac{\text{M Wt of sample} \times \text{Blank} - \text{Assay} \times \text{Normality of HCl} \times 100}{1000 \times \text{Weight of the sample}}$$

Comparison of conventional and microwave method

Sample	*% Purity*	
	Conventional method (3 hours)	*Microwave method, 100% (minutes)*
Benzyl Benzoate	51.94	71.63 (6)
Methyl Benzoate	45.72	67.18 (6)
Dibutyl phthalate	53.02	79.99 (9)
Diethyl Phthalate	51.04	67.06 (8)

Alcoholic potassium hydroxide solution (0.5 N): Dissolve potassium hydroxide pellets (1.2 g) in 95% ethanol (100 ml) and allowed to stand overnight. Filter the solution and standardize with hydrochloric acid.

Post Lab Questions:

1) Why ethanolic potassium hydroxide is used in the ester hydrolysis?
2) What is the role of heating sink?

Estimation of Carbonyl Compounds

Principle: Carbonyl compounds (aldehydes or ketones) react with hydroxylamine hydrochloride in presence of a weak base like pyridine and forms oximes. The chemical conversions of carbonyl compounds into corresponding oximes are basis for the estimation of carbonyl compounds. The liberated hydrochloric acid can be titrated with standard alkali. This reaction eliminates an equivalent amount of hydrochloric acid. The volume of hydrochloric acid is useful in the calculation. The microwave method requires heating for fewer minutes, whereas the conventional method requires heating for hours (~2 hour).

$$R-C(=O)-R^1 \xrightarrow[\text{NH}_2\text{OH. HCl, Pyridine}]{\text{MW 680 W}} R-C(=N-OH)-R^1 + HCl + H_2O$$

Carbonyl compopund → Oxime

Chemicals required

- Sample
- Hydroxylamine hydrochloride (0.5 N)
- Sodium hydroxide (ethanolic, 0.5 N)
- Hydrochloric acid (0.5 N)
- Pyridine
- Methyl orange indicator

Procedure:

1) Transfer the sample (1 g), hydroxylamine hydrochloride (50 ml, 0.5 N) into the reaction flask and add pyridine (10 ml).
2) Place the flask into the Microwave oven and switch on the magnetic stirrer.
    - In case of Microwave synthesizer, attach the reflux condenser to the flask.
    - In case of domestic microwave oven, cover the flask with a petridish loaded with ice cubes (provides condensing effect), alternatively funnel can be used. Beaker containing cold water can be used as a heating sink. It absorbs excessive energy.

3) Irradiate the reaction mixture present in reaction flask with Microwaves at 80% intensity for appropriate time, with intermittent cooling for 2 min after every minute of microwave irradiation.
4) Turn off the power supply to the Microwave oven.
5) After few minutes (allow it to cool), remove the flask from the Microwave oven.
6) Titrate the mixture with ethanolic sodium hydroxide (0.5 N) using methyl orange as an indicator.
7) Perform blank titration using hydroxylamine hydrochloride (50 ml, 0.5 N) add pyridine (10 ml) omitting the sample.
8) The percentage purity of carbonyl compounds in the sample can be calculated using the formula given below.

   % Purity =

$$\frac{\text{M Wt of sample} \times \text{Blank} - \text{Assay} \times \text{Normality of HCl} \times 100}{1000 \times \text{Weight of the sample}}$$

Comparison of conventional and microwave method

Sample	*% Purity*	
	Conventional method (2 hours)	***Microwave method, 80% (minutes)***
Acetophenone	85.49	85.80 (21)
Benzophenone	98.80	99.29 (20)
Cyclohexonone	97.23	98.21 (10)
Benzaldehyde	92.39	93.45 (6.5)
Acetaldehyde	45.84	46.59 (10)
Glucose	45.07	59.49 (10)

Post Lab Questions:

1) What are oximes?
2) What is the function of pyridine in the estimation?

EXPERIMENT 59

Assay of Riboflavin

Principle: Riboflavin (vitamin B_2) occurs in the plant (green vegetables and wheat germ) and animal kingdoms (yeast, egg yolk, milk, meat, fish). It is central component in cofactors namelyflavin mononucleotide (FMN) and flavin adenine dinucelotide (FAD). These cofactors are essential for the cellular processes such as energy metabolism. It prevents free radical damage (antioxidant) and promotes growth. It regulates blood, heart, skin, eye and hormonal functions. Riboflavin deficiency results in tongue inflammation, sore throat and, lip and mouth cracks. The aqueous solution of riboflavin shows yellowish-green fluorescence. This property is useful in the quantitative determination. Riboflavin exhibits absorbance peaks at 223, 267, 375 and 444 nm. The peaks at 267, 375 and 444 nm are stable between pH 2 to 7. The solution of the riboflavin (raw material and tablets) in presence of sodium acetate-acetic acid buffer maintains the pH. The intense peak at 267 nm is used in the assay of riboflavin raw material and 444 nm is adopted for the tablets. The content of riboflavin is determined using the Beer-Lambert's law $A = E_{(1\%)}Ct$, where 323 is the $E_{(1\%)}$, and $t = 1$ cm. The conventional heating requires 60 minutes for hydrolysis to result, but microwave method requires 4 minutes.

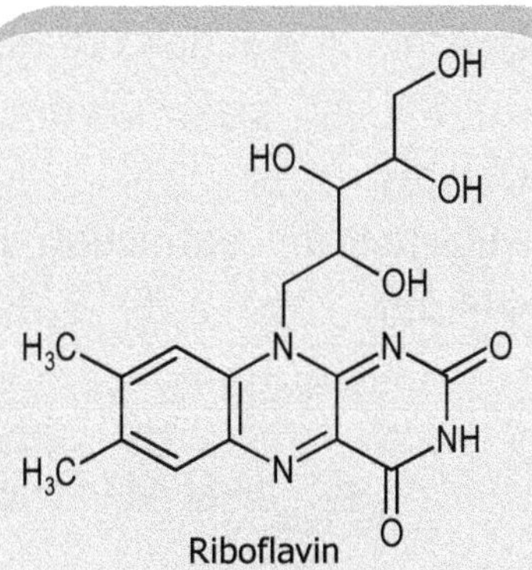

Riboflavin
(vitamin B2)

Chemicals required

- Riboflavin (raw materials/tablets)
- Glacial acetic acid
- Sodium acetate

Procedure:

1) Transfer the sample riboflavin (75 mg) into the reaction flask.
2) Dissolve it with the aid of distilled water (150 ml) and glacial acetic acid (2 ml).
3) Place the flask into the Microwave oven and switch on the magnetic stirrer.
   - In case of Microwave synthesizer, attach the reflux condenser to the flask.
   - In case of domestic microwave oven, cover the flask with a petridish loaded with ice cubes (provides condensing effect), alternatively funnel can be used. Beaker containing cold water can be used as a heating sink. It absorbs excessive energy.
4) Irradiate the reaction mixture present in reaction flask with Microwaves at 100% intensity for 4 minutes, with intermittent cooling for 2 min after every minute of microwave irradiation.
5) Turn off the power supply to the Microwave oven.
6) After few minutes (allow it to cool), remove the flask from the Microwave oven.
7) Transfer the mixture to volumetric flask (250 ml) and make up the volume with distilled water.
8) Pipette out solution (10 ml), add sodium acetate (3.5 g, 0.1 M) and make up the volume with distilled water (50 ml).
9) Measure the extinction of the resulting solution at 444 nm.

$$a = E_{1cm}^{1\%}\,ct \qquad c = a \,/\, E_{1cm}^{1\%}\,t$$

$E_{1cm}^{1\%}$ = at 444 λmax is 323; t = 1 cm

Caution: Riboflavin is light sensitive and undergoes rapid oxidation. Hence, wrap the flask with aluminum foil. The assay should be conducted in dark or lower intensity light.

Post Lab Questions:

1) Explain the term molecular extinction coefficient.
2) What are the deficiency disorders of riboflavin?
3) What is the role of sodium acetate in this estimation?

EXPERIMENT 60

Estimation of Aspirin

Principle: Aspirin (acetyl salicylic acid) is an ester of salicylic acid and acetic acid. In this titration, aspirin is heated with sodium hydroxide to completely hydrolyze to sodium salicylate and sodium acetate. The excessive (unreacted) sodium hydroxide can be estimated by titrating against standard hydrochloric acid using phenol red as an indicator (back titration). The conventional heating requires 10 minutes for hydrolysis to result, but microwave method requires 2 minutes.

$$\underset{\text{Aspirin}}{C_6H_4(OCOCH_3)(COOH)} + 2\,NaOH \xrightarrow[\text{2 min}]{\text{MW 510 W,}} \underset{\text{Salicylic acid}}{C_6H_4(OH)(COONa)} + \underset{\text{Sodium acetate}}{CH_3COONa}$$

Requirements:

- Sample
- Ethanol (95%)
- Sodium hydroxide (0.5 N)
- Hydrochloric acid (0.5 N)
- Phenol red indicator

Procedure:

1) Transfer the sample (1.5 g) into the reaction flask.
2) Dissolve the sample with the aid of ethanol (95%; 15 ml) and sodium hydroxide (50 ml, 0.5 N).
3) Place the flask into the Microwave oven and switch on the magnetic stirrer.
   - In case of Microwave synthesizer, attach the reflux condenser to the flask.
   - In case of domestic microwave oven, cover the flask with a petridish loaded with ice cubes (provides condensing effect), alternatively funnel can be used. Beaker containing cold water can be used as a heating sink. It absorbs excessive energy.

4) Irradiate the reaction mixture present in reaction flask with Microwaves at 60% intensity for appropriate time, with intermittent cooling for 2 min after every minute of microwave irradiation.
5) Turn off the power supply to the Microwave oven.
6) After few minutes (allow it to cool), remove the flask from the Microwave oven.
7) Titrate the mixture with hydrochloric acid (0.5 N) using phenol red as an indicator
8) Perform blank titration using ethanol (15 ml, 95%) and sodium hydroxide (50 ml, 0.5 N) omitting the sample.
9) The percentage purity of aspirin in the sample can be calculated using the formula given below.

   Each ml of 0.5 N sodium hydroxide is equivalent to 0.04504 g of aspirin

   % Purity =

$$\frac{0.04504 \times \text{Vol. of NaOH} \times \text{Normality of NaOH} \times 100}{\text{Actual normality of NaOH} \times \text{Weight of the sample}}$$

Post Lab Questions:

1) Exlain the principle involved in the estimation of aspirin.
2) What are the alkaline hydrolysis products of aspirin?

Estimation of Aspirin in Aspirin-Caffeine Tablets

Principle: The aspirin present in aspirin caffeine tablets can be estimated by alkaline hydrolysis with sodium citrate. The hydrolysis liberates salicylic acid and acetic acid. The liberated acid reacts with sodium hydroxide using phenolphthalein as indicator. The conventional assay requires 10 minutes. The microwave method described here reduces the duration to 2 minutes with satisfactory results.

Chemicals required

- Aspirin-caffeine tablets
- Sodium citrate
- Sodium hydroxide (0.5 N)
- Hydrochloric acid (0.5 N)
- Phenol red indicator

Procedure:

1) Transfer the tablet powder (0.7 g) and sodium citrate (2 g), into the reaction flask.
2) Dissolve the sample with the aid of ethanol (15 ml, 95%) and sodium hydroxide (50 ml, 0.5 N).
3) Place the flask into the Microwave oven and switch on the magnetic stirrer.
   - In case of Microwave synthesizer, attach the reflux condenser to the flask.
   - In case of domestic microwave oven, cover the flask with a petridish loaded with ice cubes (provides condensing effect), alternatively funnel can be used. Beaker containing cold water can be used as a heating sink. It absorbs excessive energy.
4) Irradiate the reaction mixture present in reaction flask with Microwaves at 40% intensity for 2 minutes, with intermittent cooling for 2 min after every minute of microwave irradiation.
5) Turn off the power supply to the Microwave oven.

6) After few minutes (allow it to cool), remove the flask from the Microwave oven.
7) Titrate the mixture with hydrochloric acid (0.5 N) using phenol red as an indicator
8) Perform blank titration using ethanol (95%; 15 ml) and sodium hydroxide (0.5 N; 50 ml) omitting the sample.
9) The percentage purity of aspirin in the sample can be calculated using the formula given below.

   Each ml of 0.5N sodium hydroxide is equivalent to 0.04504 g of aspirin

   % Purity =

$$\frac{0.04504 \times \text{Vol. of NaOH} \times \text{Normality of NaOH} \times 100}{\text{Actual normality of NaOH} \times \text{Weight of the sample}}$$

Post Lab Questions:

1) What are the alternate methods for simultaneous estimation of aspirin and caffeine?
2) What are the medicinal uses of aspirin?

PART - 5

NATURAL COMPONENT ISOLATION

EXPERIMENT 62

Isolation of Red dye of *Caesalpinia sappan*

Principle: Several members of the species *Caesalpinia* (Family: Caesalpiniaceae) are being used traditionally and they have a wide variety of ethnomedical properties. There are innumerable references of the use of heartwood of *Caesalpinia sappan* in the traditional medicine. It is useful in vitiated conditions of pitta, burning sensation, wounds, ulcers, leprosy, skin diseases, diarrhoea, dysentery, epilepsy, convulsions, menorrhagia, leucorrhoea, diabetes, haemoptysis, haemorrhages, stomatopathy and odontopathy. The plant is one of the ingredients of an indigenous drugs "Lukol" which is administered orally for the treatment of non-specific leucorrhea (post I.U.D.) and gave encouraging results for bleeding following IUD insertions. The wood is orange red, hard, very heavy, straight grained with a fine and even texture. As per Yunani, the wood stops bleeding from the chest and lungs, heals wounds, ulcers, improves complexion and useful in rheumatism. A decoction of the wood is a powerful emmenogogue and is used for disturbances of menstrual functions.

Procedure:

1) Transfer heartwood powder (10 g) into a Microwave flask (beaker 500 ml).
2) Add distilled water (400 ml).
3) Place the flask into the Microwave oven and switch on the magnetic stirrer.
    - In case of Microwave synthesizer, attach the reflux condenser to the flask.
    - In case of domestic microwave oven, cover the flask with a petridish loaded with ice cubes (provides condensing effect), alternatively funnel can be used. Beaker containing cold water can be used as a heating sink.
4) Irradiate the reaction mixture present in reaction flask with Microwaves at 60% intensity for 20 minutes.
5) Turn off the power supply to the Microwave oven.

6) After few minutes (allow it to cool), remove the flask from the Microwave oven.
7) Filter and Collect the filtrate and evaporate at 50 – 60 °C until dryness.
8) Weigh the red product and report its yield.

	Conventional	**Microwave (60%)**
Duration (min)	120	20

Identification Tests

1. Spray the TLC plate with ferric chloride and find out the number of phenolic compounds present in the dye.
2. **Shinoda test:** To the alcoholic solution of the red dye add magnesium metal and few drops of concentrated hydrochloric acid. The formation of red colour indicates the presence of flavonoids.

Post Lab Questions:

1) What are the therapeutic uses of *Caesalpinia sappan* ?
2) What is Lukol ?
3) What is Shinoda test ?

Isolation of Mucilage from Plants

Principle: The synthetic polymers used as pharmaceutical excipients suffer from many disadvantages such as high cost, non-biodegradability and environmental pollution. Natural polymers such as gums and mucilages are easy to isolate, purify and are non-toxic and compatible with biologic system. They are biodegradable and will not cause environmental pollution. Mucilages are polysaccharide complexes of sugar and uronic acids and are used as binding, thickening, emulsifying, suspending and stabilizing agents in pharmaceutical industries. Mucilages are isolated by soaking the plant powder for 24 h in distilled water and heating the mixture for 1 h. After filtration and addition of two volumes of ethanol to the filtrate, the mucilage gets precipitated. The conventional isolation procedure for mucilages consumes more time and yields low.

Requirements:

Seeds

[Fenugreek, *Trigonella foenum-graecum* (family Leguminosae)

- Linseed, *Linum usitatissimum* Linn (family Linaceae)
- Isphagula, *Plantogo ovata* (Plantaginaceae)
- Indian mustard, *Brassica juncea* Czern. & Coss. (family Brassicaceae) seeds

Procedure:

1) Dry the collected seeds under shade and powder in a mixer grinder.
2) Weigh 10 g of powder and soak it in distilled water (300 ml) for 24 h, in a 1000 ml beaker.
3) Place the flask into the Microwave oven and switch on the magnetic stirrer.
   - In case of Microwave synthesizer, attach the reflux condenser to the flask.
   - In case of domestic microwave oven, cover the flask with a petridish loaded with ice cubes (provides condensing effect), alternatively funnel can be used. Beaker containing cold water can be used as a heating sink.

- Black gram, *Vigna munga* Linn. Hepper (family Fabaceae) seeds]
- Ethanol

4) Irradiate the reaction mixture present in reaction flask with Microwaves at 180 W, 60% intensity for 40 minutes.
5) Turn off the power supply to the Microwave oven.
6) After few minutes (allow it to cool), remove the flask from the Microwave oven.
7) Keep it aside for 2 h for the release of mucilage into water.
8) Filter the mixture through a muslin cloth and wash the mark with hot distilled water (50 ml)
9) Squeeze well to remove the mucilage completely from the marc.
10) To the filtrate add double the volume of ethanol to precipitate the mucilage.
11) Cool and filter (or centrifuge). Dry the residue in an incubator at 37° C. Powder, weigh and report the yield.

Comparison of conventional and microwave method

	Yield (g)	
Plant source	**Conventional Method (1 h)**	**Microwave method (180 W; 40 min)**
Fenugreek	2.642	3.424
Linseed	1.138	1.691
Isphaghula	1.864	4.756
Indian mustard	0.936	1.610
Black gram	3.152	3.946

Identification Tests

Test the mucilage for the following tests to confirm their identity.

- To a solution of the mucilage add a solution of ruthenium red: mucilage stains pink.
- To a solution of the mucilage add a solution of corallin soda and sodium bicarbonate solution (25%): mucilage stain pink.

Post Lab Questions:

1) What are the uses of polymers?
2) List and explain about the identification test for mucilage.

Isolation of Pectins from Plants

Principle: Pectin (polysaccharide) contains polygalacturonic acids in which some of the carbonyl groups are present in methyl esters along with araban and galactan. These occur in the middle lamellae of cell walls and are abundant in fruits namely apples, papaya, guava, mango and oranges. Pectin is isolated by heating the raw materials with water adjusted to pH 3.5-4.0 with tartaric acid and pouring the filtrate into three volumes of acetone. The conventional isolation procedure required more heating time. In the present experiment, several pectin's were isolated in higher yields using a microwave technique.

Pectin, by addition of water forms the lumps which on heating goes to solution. This property widens its applications in food, pharmaceutical and several other industries. Pectin is used as a jelling agent in making various jellies, as a thickening agent in chocolate syrups and as a preservative in food products. Medically, it is used as an intestinal tract regular and for intravenous treatment of shock. D-galacturonic acid prepared from pectin is used to synthesize vitamin C. Salts of pectin are used as aids in spray drying, paper coating and as substitute for agar agar in bacteriological media.

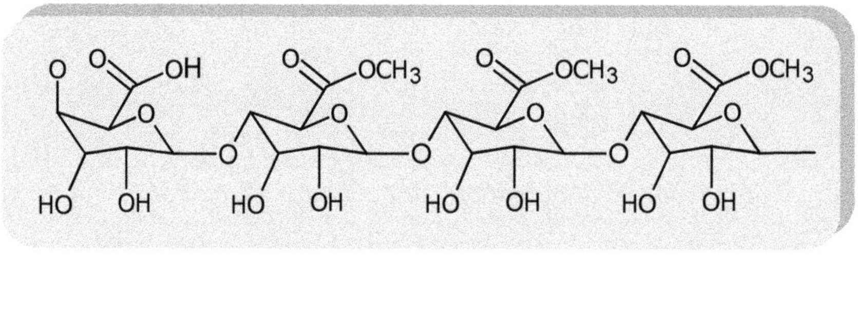

Requirements:

Plants

Orange,

Citrus reticulata, Blanco (family Rutaceae), dried peels

- Guava, *Psidium guajara*, Linn (family Myrtaceae), pericarp
- Mango, *Mangifera indica*, Linn (family Anacar-diaceae), pericarp
- Ethanol

Procedure:

1) Mince the pericarp of the collected material (50 g) in a mixer grinder and soak it in distilled water (200 ml) for 2 h, in a 1000 ml beaker.
2) Adjust the pH of the solution to 4.5 with a solution of tartaric acid (10%).
3) Place the flask into the Microwave oven and switch on the magnetic stirrer.
   - In case of Microwave synthesizer, attach the reflux condenser to the flask.
   - In case of domestic microwave oven, cover the flask with a petridish loaded with ice cubes (provides condensing effect), alternatively funnel can be used. Beaker containing cold water can be used as a heating sink.
4) Irradiate the reaction mixture present in reaction flask with Microwaves at 510 W, 60% intensity for 20 minutes.
5) Turn off the power supply to the Microwave oven.
6) After few minutes (allow it to cool), remove the flask from the Microwave oven.
7) Filter while hot through a muslin cloth.
8) Cool the filtrate and pour into a beaker containing acetone (600 ml) with continuous stirring to precipitate out pectin.
9) Filter (or centrifuge), wash with acetone to make it free from acidic impurities.
10) Dry the residue in an incubator at 37° C. Powder, weigh and report the yield.

Comparative durations:

Plant source	Yield (g)	
	Conventional Method (1 h)	**Microwave method (350 W; 20 min)**
Dried orange peel	0.432	0.658
Mango	0.579	0.749
Guava	0.349	1.016

Identification Tests

Test the pectin for the following tests to confirm their identity.

- A 10% solution of the pectin forms a stiff gel on cooling
- To 5 ml of 1% solution, add 1 ml of 2% solution of potassium hydroxide and set aside at room temperature for 15 min. A transparent gel or semi gel forms.

Acidify the gel with dil HCl and shake well. A voluminous, colourless, gelatinous precipitate forms, which when boil becomes white and flocculated.

Post Lab Questions:

1) What are pectins?
2) List pharmaceutical uses of pectins.

EXPERIMENT 65

Isolation of Phenolic Compounds from Grape Seed

Principle: Phenols and polyphenols in wine, fruits and vegetables have been mainly attributed to their antioxidant properties, which prevent free radical mediated cytotoxicity and lipid peroxidation and oxidation of low density lipoproteins (LDL). Grape seeds have been confirmed as a rich source procyanidins or condensed tannins and other phenolic compounds. Grape seed procyanidins extracts at 100 mg/L concentration exhibited 78 – 81 % inhibition of superoxide anion and hydroxyl radical, while vitamin C showed 12 – 19% under similar conditions. Due to this potent antioxidant activity, several formulations containing them are being marketed in USA and several other countries. Extraction of grape seed polyphenols is usually performed under low temperatures, protected against oxidation with CO_2 or ascorbic acid and using methanol as the solvent. In the present experiment, the grape seed poly phenols are isolated under microwave irradiation.

Requirements:

- Grapes
- Methanol
- Petroleum ether

Procedure:

1) Place the dried grape seeds (15 g) into a round bottom flask and add petroleum ether (50 ml) and reflux for 30 minutes on a water bath.
2) Filter, collect the residue, dry.
3) Place the dried and defatted grape seed powder in a microwave flask (beaker, 250 ml) and add 90% methanol (150 ml).
4) Place the flask into the Microwave oven and switch on the magnetic stirrer.

- In case of Microwave synthesizer, attach the reflux condenser to the flask.
- In case of domestic microwave oven, cover the flask with a petridish loaded with ice cubes (provides condensing effect), alternatively funnel can be used. Beaker containing cold water can be used as a heating sink.

5) Irradiate the reaction mixture present in reaction flask with Microwaves at 160 W, 60% intensity for 2 minutes.
6) Turn off the power supply to the Microwave oven.
7) After few minutes (allow it to cool), remove the flask from the Microwave oven.
8) Filter and evaporate the methanol (preferably in a rotary evaporator) at 40° C, weigh the residue, and report the yield.

Determination of total phenol content: Total phenol content can be determined through Folin-Ciocalteu method.

Principle: The total phenol content determination is based on the oxidation of phenolic groups with Folin-Ciocalteu reagent. The regent contains phosphomolybdic and phosphotungstic acids. Folin-Ciocalteu reagent oxidizes the phenols and generates a green blue complex. The absorption of the complex can be measured at 750 nm.

Requirements:

- Folin-iocalteu reagent
- Sodium carbonate

Procedure:

1) To 0.4 ml of the sample solution, add 2 ml of Folin-Ciocalteu reagent (1:10 in distilled water)
2) Add sodium carbonate (2 ml, 15%) in distilled water.
3) Mix the contents well by shaking and keep it aside for 2 h to complete the colour development.

4) Measure the absorbance of the green coloured complex at 750 nm.
5) Calibration curve:
   Standard: Gallic acid monohydrate (1-10 μg/ml)

Observations:

1. Weight of plant material = g
2. Yield of phenolic residue = g
3. Percentage yield = %
4. Total phenol content = mg/g of gallic acid

Post Lab Questions:

1) Explain about the Folin-Ciocalteu method of determination
2) What are specific conditions required for the extraction of grape seed polyphenols.

EXPERIMENT 66

Isolation of Piperine from Pepper

Principle: Piperine is a pungent alkaloid present in the *Piper nigrum* and *Piper longum*. The chemical name of piperine is trans, trans-5-(3,4 methylenedi oxyphenyl)-2,4-pentadienoic acid piperidide. The major therapeutic/physiological functions of piperine include antioxidant, neuroprotective, anticancer and anti-inflammatory effects. It also used a major flavoring agent.

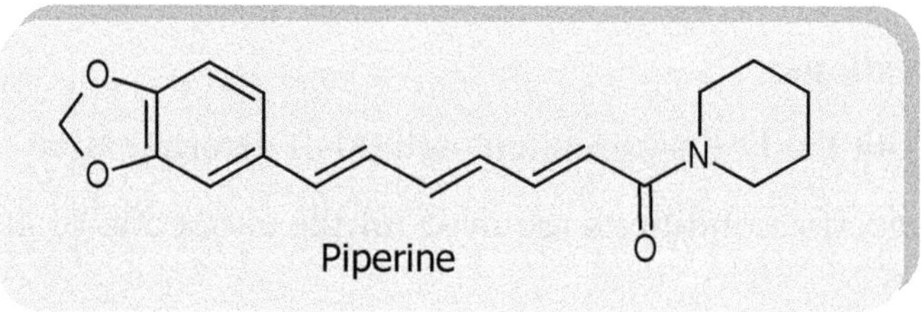

Piperine

Chemicals required

- Pepper
- Dichloromethane (DCM)
- Diethyl ether
- Ethanol

Procedure:

1) Transfer powdered pepper (2 g) into reaction flask (100 ml) and dissolve with dichloromethane (15 ml).
2) Place the flask into the Microwave oven and switch on the magnetic stirrer.
    - In case of Microwave synthesizer, attach the reflux condenser to the flask
    - In case of domestic microwave oven, cover the flask with a petridish loaded with ice cubes (provides condensing effect), alternatively funnel can be used. Baker containing cold water can be used as a heating sink. It absorbs excessive energy.

3) Irradiate the reaction mixture present in reaction flask with Microwaves at 100% intensity for 7 min, with intermittent cooling for 1 min after every 2 min of microwave irradiation.
4) Turn off the power supply to the Microwave oven.
5) After few minutes (allow it to cool), remove the flask from the Microwave oven.
6) Place the flask in an ice bath and add diethyl ether (2 ml) to precipitate the solid.
7) Dissolve the solid with ethanol (2 ml)
8) Add the ethanolic extract into beaker containing ethanolic potassium hydroxide (10%, 2 ml) to precipitate the acidic resins.
9) Filter to remove the solid and allow it cool to produce yellow precipitate of piperine.
10) Filter and wash the crystals with cold water.

Comparison of conventional and microwave method

	Conventional	***Microwave***
Heating time	2 h	7 min 100% intensity (850 W)
Reaction temperature	45 °C	45 °C
Colour	Yellow crystals	
Melting point	128-130 °C	
Yield	94%	99%

Post Lab Questions:

1) Name the sources of piperine.
2) List the therapeutic uses of piperine.
3) Why potassium hydroxide is added during the isolation?

EXPERIMENT 67

Isolation of Caffeine from Tea

Principle: Caffeine is a xanthine class of alkaloid found in tea, coffee and chocolate. It is chemically 1,3,7-trimethyl-3,7-dihydro-1H-purine-2,6-dione. It is known for the stimulant effect on central nervous system (CNS), heart, kidney and lungs. The excessive use of this leads to headache, nervousness, insomnia and addiction. Caffeine can be isolated from the tea leaves using a base. The bases such as calcium carbonate and sodium carbonate helps in converting tannins and gallic acid into their salt (calcium or sodium salts) form (insoluble in organic solvent). This chemical modification assists in the isolation of caffeine without any interference.

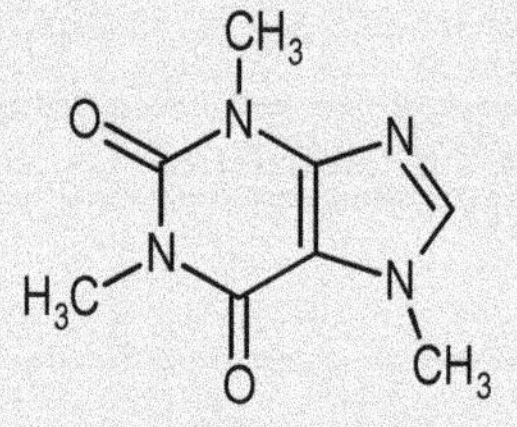

Caffeine

Chemicals required

- Tea dust
- Potassium carbonate (or sodium carbonate)
- Dichloromethane (or Chloroform)
- Distilled water

Procedure:

1) Transfer tea powder (10 g) into a flask (beaker 500 ml) and add distilled water (100 ml).
2) Shake the flask (or beaker) for about 1.5 hour at room temperature.
3) Transfer the mixture into microwave flask.
4) Place the flask into the Microwave oven and switch on the magnetic stirrer.

- In case of Microwave synthesizer, attach the reflux condenser to the flask.
- In case of domestic microwave oven, cover the flask with a petridish loaded with ice cubes (provides condensing effect), alternatively funnel can be used. Beaker containing cold water can be used as a heating sink.

5) Irradiate the reaction mixture present in reaction flask with Microwaves at 850 W intensity for 10 min with intermittent cooling for 1 min after every 2 min of MW irradiation.
6) Turn off the power supply to the Microwave oven.
7) After few minutes (allow it to cool), remove the flask from the Microwave oven.
8) Filter at the pump, transfer to separating funnel and extract the mixture with 15ml of dichloromethane (or chloroform) for two times.
9) Collect and concentrate the organic layer to precipitate the caffeine or evaporate organic layer on petriplate at room temperature to get white crystals of Caffeine.

Note: Repeat the extraction several times to isolate the caffeine in good yields.

Comparison of conventional and microwave method

	Conventional	***Microwave***
Heating time	90 min	10 min 100% intensity (850 W)
Reaction temperature	80-90 °C	100 °C
Colour	White crystals	
Melting point	227-228 °C	
Yield	90%	98%

Post Lab Questions:

1) Name the sources of caffeine.
2) List the therapeutic uses of caffeine.
3) Why potassium carbonate is used during the isolation of caffeine?

INDEX

www.ingramcontent.com/pod-product-compliance
Ingram Content Group UK Ltd.
Pitfield, Milton Keynes, MK11 3LW, UK
UKHW052231270726
14060UKWH00005B/700

9 789389 974959